Drainage Windmills on the Broads

Patrick Taylor

POLYSTAR PRESS

ISBN 978 1 907154 61 4

Drainage Windmills on the Broads

Published by Polystar Press

62 Angel Street 277 Cavendish Street
Hadleigh Ipswich
Suffolk Suffolk
IP7 5EY IP3 8BQ
(01473) 824896 (01473) 434604
polystar@btinternet.com polystar@ntlworld.com

ISBN 978 1 907154 61 4

All rights reserved.
This book is protected by copyright.
No part of it may be reproduced, stored in a
retrieval system, or transmitted, in any form
or by any means, electronic, mechanical,
photocopying, recording or otherwise, without
the written permission of the author or publisher.

Every attempt has been made to trace accurate
ownership of copyrighted material in this book.
Errors and omissions will be corrected in
subsequent editions, provided that
notification is sent to the publisher.

© Patrick Taylor 2017

Typeset by nattygrafix

Printed by
R Booth Ltd, The Praze, Penryn

Contents

section		page
0.0	Introduction	1
1.0	**The Drainage Story**	
1.1	Topography and Prehistory	2
1.2	Peat Digging	4
1.3	Watermills	5
1.4	Land Drainage	6
1.5	Drainage Mills	8
1.6	Variety and Survival	10
1.7	Map Sources	13
1.8	Documentary Sources	14
1.9	The Present Work	17
2.0	**A Gazetteer of Mills and Sites**	
2.1	River Thurne	18-33
2.2	River Ant	34-46
2.3	River Bure: Upper	47-51
	River Bure: Middle	52-63
	River Bure: Lower	64-73
2.4	River Yare: Upper	74-79
	River Yare: Middle	80-87
	River Yare: Lower	88-99
2.5	River Waveney: Upper	100-109
	River Waveney: Lower	110-115
3.0	**End Papers**	
3.1	Summary	117
3.2	References and Bibliography	118
3.3	Acknowledgements	119

In memory of Dil Andrews (1953-2017)

She enjoyed her boating

0.0 Introduction

Two Norfolk Broads holidays, taken in May 1990 and June 1994, provided a legacy of colour slides of the mills of this area, mostly taken from a viewpoint afloat. Until recently these were hidden in a drawer, but retirement brought with it a digital slide converter and time in which to pursue the subject. There did not seem to be a comprehensive guide to the subject and what was available was dated, sometimes inaccurate and poorly presented for the person wanting to visit these buildings whilst on a similar boating holiday.

This book presents all the drainage windmills that can be seen, along with the sites from which they have vanished, hopefully in a comprehensive and useful manner. It deals with the area river by river, and within that mill by mill as one passes downstream on each waterway, rather than in the alphabetical order adopted by previous authors. It also deals with the entire river basin here including the mills on the southern side of the Waveney in the seemingly foreign land of Suffolk.

Indeed the book has largely been written in Suffolk, but has required a number of walks to be undertaken in Norfolk along parts of each river in turn to refresh the information collected on the original holidays, a method more useful for reaching some of the mills not accessible by boat.

The author has thus added some 50 miles of walking to his previous 150 miles or so of cruising on the original holidays. Long distance footpaths in the form of the Angles Way along the Waveney, the Wherryman's Way along the Yare and the Weaver's Way along parts of the Bure and Thurne now provide a useful alternative way to visit this area.

The author's 1990's slides provide a useful baseline, but where a mill has undergone significant restoration or repair since the original survey, newer pictures have been included where they rightly belong in the gazetteer. The older pictures illustrating the degree of change are then also included wherever possible.

Limpenhoe Marshes, River Yare

1.1 Topography and Prehistory

The Norfolk Broads is a term now used to describe an area of low lying wetland mainly in east Norfolk, but bordering Suffolk to the south. It now comprises a single drainage bowl where the Rivers Bure, Yare and Waveney deliver their waters into the North Sea at Great Yarmouth. This delivery is believed to have originally been fourfold in nature.

The River Thurne that now runs away from the coast westwards to join the Bure is thought to have once taken the headwaters of the Bure eastwards in the opposite direction along the course of the Hundred Stream, reaching the coast between Horsey and Somerton. The lower waters of the River Bure are similarly thought to have once gone direct into the sea near Caister, rather than via Breydon Water and Great Yarmouth as they do today. The River Yare itself went to Great Yarmouth as now, but originally the River Waveney met the sea separately further south at Lowestoft in north Suffolk.

The rivers remain tidal in their lower reaches and were the various river banks, coastal dunes and modern defences not there, the coastline would by now be considerably nearer Norwich. In behind Great Yarmouth the Rivers Yare and Waveney are conjoined several miles inland by the New Cut.

This was originally opened in 1833 as part of the Norwich and Lowestoft Navigation which was planned to allow Norwich access to the sea outside of the control of the competing port of Great Yarmouth. It now conveniently links the two rivers, allowing navigation between these two waterways without the need to go further downstream and then back again via the head of Breydon Water.

Our Palaeolithic forebears left hand axes and fossil footprints, as found below the cliffs near Happisburgh, and later Mesolithic remains have been widely dredged off the north Norfolk coast. During the melt following the last Ice Age, the sea level rose and by about 6,000 BC the land bridge connecting Britain to Europe had at last been broken.

The Wherry 'Albion'

photo: mark page 2011

From about 4,000 BC Neolithic man has also left his mark, mostly along the river valleys in the east of England. His timber circles, made of wood in the absence of suitable stone, have been discovered alongside the upper reaches of the River Yare at Arminghall near Norwich (1936) and the River Waveney at Flixton above Bungay (1996). These are believed to have been ceremonial in purpose and the latter has an impressive alignment towards the northernmost setting point of the moon.

By Roman times it seems the sea reached its highest level as the area of Breydon Water was a 'great estuary' adjoined on its southern bank by the large stone-built fort 'Gariannonum', now known as Burgh Castle. Opposite on the north bank near Caister, there was a walled Roman town matched by another further upstream at Caistor St Edmund south of Norwich, probably at the time the limit of useful navigation.

Nowadays the estuary is just Breydon Water and the whole area around the lower reaches of all three main rivers is essentially grazing marsh, described in the Domesday Book as meadow, divided up amongst various village ownerships. These marshes were often at some distance from the village concerned and sometimes connected by droving routes allowing the easy transfer of stock.

One of the largest and least accessible of the areas of marshland is The Island, in the triangle formed by the New Cut and the Rivers Yare and Waveney, with a quarter mile of road clipping its southern apex and a perimeter footpath along most of its sea walls. Here can be found marshes named after Chedgrave and Langley, both villages some six miles to the west.

Similarly further north between the rivers Bure and Yare a larger expanse between the two railway routes into Great Yarmouth contains marshes named after Acle and South Walsham, respectively five and seven miles to the north-west.

Burgh Castle

1.2 Peat Digging

Apart from Breydon Water, the broads themselves are also wider expanses of water, but further upstream, either alongside or sometimes astride the natural courses of the rivers in their middle reaches. Originally thought natural, in 1960 they were shown by a multi-disciplinary team to be the remains of extensive historic peat digging enterprises during early Medieval times. Estimates have been made that some 900 million cubic feet of peat has been extracted over the centuries, much of it taken to Norwich.

It now seems likely that the peat diggings originated with a 9th Century influx of Danish settlers, particularly in the hundred known as Flegg, north-west of Great Yarmouth. Here there is a concentration of village names ending in "by" and back in Denmark the use of peat as a fuel is well documented right back into the Iron Age. Flegg was originally a peninsula between the River Yare and the Hundred Stream, where the Rivers Bure, Ant and Thurne originally met the sea further north.

The peat diggings were investigated using core samples and found to be flat bottomed vertically sided pits with some un-dug peninsulas or balks between them, usually along parish boundaries or lines of ownership change. The pits are most notable for their depth as opposed to their extent, presumably a result of the best quality peat being the deeper, the wider expanse of lower quality, less valuable peat not being worthwhile exploiting in comparison.

With worsening climate and rising sea levels during the 14th Century, keeping the adjoining rivers at bay gradually became impossible, so that by the 16th Century, maps show more or less the extent of water we see today, with large expanses either alongside or totally incorporating the routes of the rivers.

No longer viable as peat workings, this led to the broads being used in later times as fisheries. Ongoing sea level rise also led to the need for the adjoining areas of grazing marsh to be managed and drained adequately.

Boom Tower, Carrow Bridge, Norwich

photo: polystar 2016

1.3 Watermills

Mills for the grinding of corn are documented as far back as in the Domesday Book of 1086. They occurred across the whole of central Norfolk, with noticeably fewer examples in the Fens to the west and the Broads to the east. This distribution of what were at that time mostly watermills in the river valleys (although there may have been some driven by animal muscle power) is largely a result of Norfolk's geology.

Basically a thousand foot or so thick bed of chalk underlies both Suffolk and Norfolk, dipping from a low scarp in the west (running from Hunstanton's cliffs south to the Gogmagog Hills near Cambridge) to disappear beneath the North Sea in the east. This chalk is overlain with rich boulder clays of glacial origin in the centres of both counties, but these clays are replaced by sandier crag deposits in the east.

The central belt is thus both the best area for growing corn and the best provided with narrow river valleys where watermills can work off a good head of water. The Fens and the Broads in contrast were less good for growing corn and are characterised by wider slow flowing rivers in their lower reaches. Here a traditional watermill would not work, as the only head of water is between the river and the adjoining marsh.

It is believed that windmills themselves were not introduced into England until the century after Domesday, and at first they were probably used to fill in the gaps in the distribution of corn grinding mills, on higher ground where there was stronger wind and no watermill competition.

Both watermills and windmills were developed over the centuries to perform other tasks that required power such as sawing, bone crushing, cement grinding and various tasks in the cloth industry such as fulling. The pumping of water became a particular speciality of the windmill.

St Mary's Church, Fishley

1.4 Land Drainage

Windmills are thought to have been first used for drainage in Norfolk in the Fenlands of the west, where massive undertakings such as the Old Bedford River of 1631, or its later parallel the New Bedford River of 1651, provided a more direct route for water removal from the marshland, by-passing the old and winding River Great Ouse on its way to Kings Lynn and eventually its outfall in The Wash.

The waters here were pumped by windmills from the surrounding marshes into the new waterways, sometimes in two lifts if the height demanded. In between the two new 'rivers', over half a mile apart, were 'The Washes', effectively a large storage reservoir for excess floodwater, often needed in the winter months.

It is thought these mills were mostly smock mills of Dutch design, brought in by the engineers such as Cornelius Vermuyden employed to implement these huge schemes that created new farmland out of the marshes. Although many of these mills were shown on Faden's map of 1797, most were replaced in the 19th Century with steam engines, which were superseded in their turn in the 20th by diesel or electric pumps.

Parts of two of these smock mills at Nordelf (TF 561003 and 561009) survive today. They are octagonal in plan, built with a battered brickwork two storey base, originally capped with a tapering timber-framed and timber-clad tower that carried the sails.

Now truncated with the brick bases converted to houses, they do bring to mind the 'Dutch Cottages' of early 17th Century date to be found on Canvey Island in Essex, where Dutch engineers also worked (see author's *'Toll-houses of Essex'*, 2010).

photo: arthur c smith 1988

Converted Smock Mill, Nordelf, Norfolk

The drainage of the Broads in the east seems to have followed a similar if slightly later adoption of wind power using similar smock mills with scoop wheels at first. A small number of other types such as trestle mills or hollow post mills were also used, however the majority are of a more local design that evolved using brick towers with boat-shaped caps, that is fairly distinctive in east Norfolk.

These eastern drainage mills seem to have survived in far greater numbers and some were built new or converted to use turbines to deliver the water into the rivers. The schemes here seem a little more piecemeal, draining one marsh belonging to a single parish at any one time, but eventually achieving a comprehensive result, much of which fortunately remains.

Typically the settlement pattern within a parish would comprise a cluster of houses, farmsteads and a manor house or two with attendant church, all situated on higher ground. The lower level wetter land adjoining the rivers would be grazing meadow and the river itself almost universally a parish boundary, as it was generally a barrier that was uncrossed except at a few bridging points. Some parishes such as Catfield, had two waterfronts abutting adjoining river basins.

Each parish thus had its own local grazing marsh, and many had further areas of parish 'detached' and further down the river on the spare land of the wide estuarine flood plains, where settlement was not feasible.

The grazing marshes themselves were generally criss-crossed by a network of drainage ditches, that fed into a number of larger main drains running down towards the river. These generally terminated in a semi-circular ditch abutting the river bank or 'wall' as it was known. The drainage mills were then built on the river bank within the semi-circle, pumping water from it into the river itself.

Former Cattle Pound, Somerleyton

1.5 Drainage Mills

The earliest mills, as we have already noted, were of the 'smock' type comprising a timber-framed, usually octagonal structure on a brick base, clad with timber boarding. The later drainage mills, as became the norm in the 19th Century, were 'tower' mills, circular in plan, with the sails mounted onto a rotating cap atop a red brick tapering tower, sometimes tarred.

Round towers of other kinds are well known elsewhere in Norfolk from medieval times onwards, either as defensive towers in Norwich's city walls or the round towers attached to so many Norfolk churches, particularly south and east of Norwich itself.

To be effective at catching the wind head on, the entire cap was mounted on rollers running on a curb in early examples, the whole thing being manually turned using a long tailpole for adjustment that almost reached the ground behind the cap. In later times mills were usually winded (not wound!) by fantails that automatically adjusted the cap's direction eliminating the need for a tailpole. These fantails were small sets of usually eight vanes set perpendicularly behind the cap and rotated it using gearing connected to the curb. If the wind changed direction the fantail would rotate driven by the crosswind until such time as the cap had moved to eliminate it.

The earliest sails, of the 'common' type, were just that, simply cloth sails such as a sailmaker might make for a boat, set by hand onto the individual stocks in turn, these last extending very nearly to the ground to facilitate this.

With the introduction of 'patent' sails after their invention in 1807, mills were built taller to better take advantage of the wind keeping the sails well clear of the ground. The patent sails usually comprised double rows of timber shutters that could be adjusted whilst still running by 'striking gear' to spill the wind during heavier weather, rather than needing reefing as would a common sail.

Scoop Wheel Gearing, Polkey's Mill

The two timber stocks, perpendicularly crossed at their centres supporting the sails, were mounted onto a nearly horizontal windshaft, timber in the earliest examples, sometimes laminated for greater strength, but later on mostly cast iron. The windshaft was situated mostly within the cap, near the centre of which it terminated in the brakewheel, basically a large cog that meshed with another, the wallower, mounted at the top of a vertical drive shaft, which then conveyed the power down to ground level. Page 16 shows a diagram of mill headgear workings using an example from Suffolk County, Long Island, New York, albeit a corn mill and foreign, but not untypical and in this case particularly well drawn.

From here onwards the machinery within a drainage mill was much simpler than that in a corn grinding mill where drives would be taken off the main one to turn the grindstones and run hoists etc. The inside of a drainage mill was still floored out into separate storeys joined by stairs and the space was sometimes used to live in, but the single vertical drive shaft simply passed down through each floor.

At its base, further gearing connected the vertical drive to a horizontal shaft on which was mounted, usually externally, the scoop wheel that lifted the water from the drainage dyke up to the river. The sides of some brick towers are flattened to accommodate this. In later mills the drive was taken to a turbine pump, which although more efficient, was less reliable in higher winds. With the greater efficiency a turbine pump offered, there was a tendency in later times to once again build smaller mills, often reverting to the timber or metal framework of a 'trestle' type mill, clad with boarding or sometimes left as an open skeleton.

Overall then we see that no particular mill is typical, each a product of its own time and site, each taking on board new techniques as they became viable in a series of upgrades over the years.

photo: polystar 1994 S22

Polkey's Mill Y24 (p.89)

1.6 Variety and Survival

The survival of such numbers of drainage mills as are found on the Broads should not be a reason for complacency as without 'beneficial use' in their original guise as working pumps, these buildings remain very much 'at risk' and comprise a mixture of those carefully restored, those converted to domestic use (often with much loss of historic fabric) and the remainder empty and derelict awaiting one of those fates.

It must have been the relative remoteness and difficulty of access to these sites that kept free wind power as the preferred option for carrying out this work of drainage compared to steam power or eventually diesel engines or electricity as took over in the western Fenland areas.

The building of a complete new wind powered drainage mill at Martham Level as late as 1908 is testament to the wind being the preferred power source in the more remote areas of the Broads right into the 20th Century. Many other mills were rebuilt at this time, sometimes in heightened form and sometimes with new turbine pumps replacing earlier scoop wheels.

However, coal certainly was available at West Caister just two miles up river from Great Yarmouth, where the 1843 Tithe Map has a circular plot (368) listed in the Apportionment as 'Steam Engine Yard'. Further upstream on the River Bure, Tall Mill at Upton was augmented c.1840 by a steam engine in a shed, the coal for which must have come up river too.

Coal itself would have been a commonplace cargo up the Rivers Waveney and Yare heading to Beccles and Norwich respectively, the trade having its roots back in the days when peat was the fuel of choice. Coal from Lowestoft meant that by 1884, many upper Waveney and upper Yare valley sites were converted to or augmented by steam. This change continued with the introduction of diesel engines and then electricity in its turn used to pump the marshes dry, so that today with the nearby offshore wind farms we are now nearer the original wind-powered pumping than for many a year.

Steam and wind power together at Upton

photo: evelyn simak 2009

Hardley Mill: Newly restored brakewheel within boat-shaped cap

photo: ashley dace 2010

Womack Water Mill: derelict remains of drive shaft and wallower

Hardley Mill: Newly restored wallower at top of drive shaft

1.7 Map Sources

Faden's 1797 map of Norfolk provides a good starting point, but predates many mills. The oldest maps that provide comprehensive coverage of the area date from c.1840. Firstly we have the early 1" to a mile Ordnance Survey maps, surveyed just prior to that date, which show many mill symbols X alongside the occasional reference to a 'Marsh Mill' or some such. Copies of these published by David & Charles c.1970 have been used here, which comprise base maps from surveys of 1837 or 1838, updated later (but not by resurvey) to include more recent additions such as railway lines.

Secondly we have the Tithe Maps, which mostly date from 1839 to 1844, and cover the area parish by parish. Unfortunately there are areas for which this source is incomplete or not available, and sometimes the detached portions of parishes remain elusive. The Tithe Maps also involved a number of different surveyors so that the information can be variable in quality. Some clearly show mill sites with a mill symbol X where a drainage mill existed at the time, whilst others just show the plan of a small circular building, both usually with an adjoining plot reference number. These numbers are then sought in the Tithe Apportionment, generally produced a year or two later and which usually refer to 'Mill & Yard' or 'Mill Mount', or if the number relates to an adjoining plot of land we might get 'Mill Meadow' or 'Mill Pightle', proof of a mill nonetheless. These two early map sources usually concur and provide good evidence for a wind powered mill at many a site.

The next useful batch of information comes from the 6" and 25" to a mile OS maps dating from c.1884, which show much more detail. Here a drainage mill would most usefully appear as 'Windmill (Pumping)' or often as 'Draining Pump', sometimes helpfully qualified by '(Wind)'. Here the shape of the plan form can be very useful, the circular ones wind-powered, the rectangular ones engine driven. Some sites show one building of each type and the text 'Draining Pumps' in the plural indicative of an auxiliary engine helping in times of low wind speed.

Hardley Mill Y12 (p.83)

1.8 Documentary Sources

The next comprehensive source of information on this subject is documentary and comprises Rex Wailes' pioneering paper presented to the Newcomen Society in 1956. His Appendix 1 provides a very useful list of sites (the entries of which, covering the area here considered, the present author has for easy reference simply numbered from 1 to 110).

According to his Acknowledgements, the information in the paper had been collected from a number of sources including several old millwrights and a number of early amateur enthusiasts. He discusses the difficulty in collecting and verifying such information, the variety of names that some mills are known by, which can confuse identification, and he does apologise for any inaccuracies, saying 'I fear that error may have crept in'.

Indeed it did creep in and we find some sites appearing twice in the list, which at least is better than a non-appearance. This may be due to different generations of millwrights having recorded working on a mill, such as at initial build and later at a heightening or turbine installation. Second time around there may well have been a different marshman operating it and hence a different name attached.

Although Wailes' list is alphabetical, it is not done by parish and leads to there being several entries under L or W for instance, where sites in various parishes are listed as above or below Ludham or Wayford Bridges. Fortunately these give accurate directions from the bridge in question to the site, so that the sometimes missing or inaccurate map references he gives do not matter.

Double entries still confuse us today in a different way as many online listings include separate entries for distinct mills, but illustrated by photos of the same mill. The present author is also wary that error may creep in, but will endeavour to sort most of this out satisfactorily.

Scoop Wheel, Lockgate Mill

Fortunately Arthur C Smith's publication in 1978 builds on Wailes' excellent example and goes some way to sorting out many of the anomalies. He provides a comprehensive photographic survey of what remained of any substance at that time and his alphabetical parish by parish listing with details of each mill in turn is easier to use than Wailes' list, although its sideways presentation within the document does impede a little.

He clarifies where necessary anything Wailes had given a different name to and also provides a short list entitled 'Some Old Sites, mostly containing no remains', but then blots his copybook by including the very substantial Neave's Mill on the River Ant amongst these, whilst illustrating and describing it earlier in the document, but at an incorrect map reference.

One further set of documentary sources that deserve mention are the List Descriptions produced for English Heritage's listing branch, mostly in the same decade as Smith's work. The present author has sometimes in the past found such descriptions unreliable (when dealing with toll-houses) and has noticed that those dealing with mills do on occasion re-iterate without question information given by Wailes in his work, which as we have found is not always totally accurate.

The listings do however provide details of the intimate inner workings of many mills, too complex for inclusion here, but certainly worthy of investigation if the reader desires more information.

Knowing what we have left of these remains of an earlier age is paramount in determining what will happen next to them. Understanding is the first step towards conservation and it is hoped the information presented here will help secure the future of many more mills.

It is also hoped that the ordering of the information using a river by river numbering system will form a good and reliable basis for all future reference.

photo: polystar 1994 S21

Cadge's Mill Y26 (p.91)

15

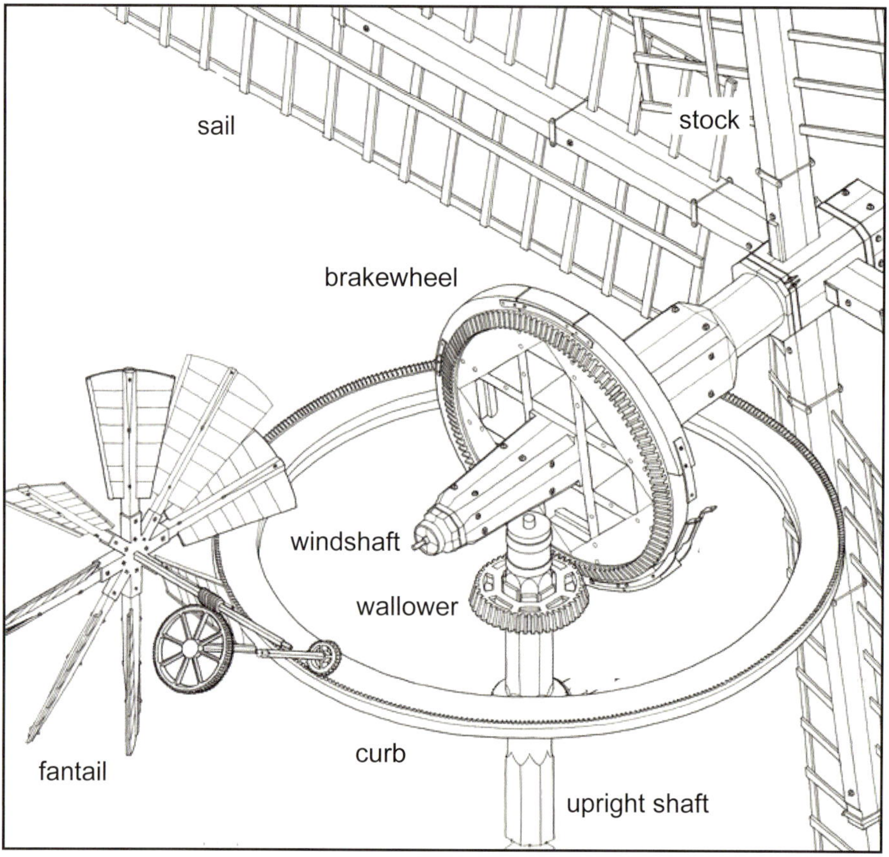

Not untypical Corn Mill headgear: Beebe Mill, Suffolk County, Long Island
drawing: long island wind and tide mill survey 1976

photo: chris allen 2014

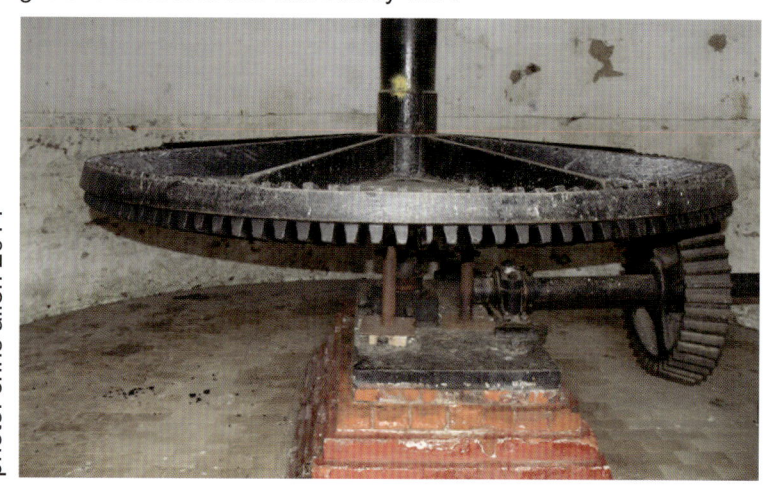

Base of upright shaft with gearing to turbine pump drive,
Stracey Arms Mill B24 (p.65)

1.9 The Present Work

The present work hopefully builds on all the invaluable earlier work of Wailes and Smith and includes the author's own photographic surveys undertaken in May 1990 (north area) and June 1994 (south area). Here on two Norfolk Broads holidays, with the kind co-operation of the author's then partner and her daughter, a set of photographs was obtained showing as many mill remains as might be seen from the water and that could be navigated past without sudden changes of direction that might propel the photographer into the drink. Whilst these photographs may not be comprehensive, they are here augmented by newer ones for the mills that these holidays missed or where substantial changes have taken place in the last twenty years.

In addition to the windmills that still present us with remains (shown as ● on the maps) and thus a photograph, this book attempts to document all the sites where windmills (shown as O) or pumps (not shown) have now gone, largely based on the early map evidence. Those shown only by Faden but not on any later 'early' maps are shown as F. The lost sites are presented in a number of text boxes within which their map references are given in brackets. The arrangement now adopted for the gazetteer that follows is not alphabetical, but runs mill by mill downstream along each river in turn.

It starts with the northern-most nearest the North Sea, the River Thurne and then runs anti-clockwise and roughly southwards through the Rivers Ant, Bure, Yare and Waveney in turn, this last forming the border with Suffolk along most of its length. It is hoped that such an arrangement will be most useful to those now taking such holidays, providing a continuity of reference as one progresses along any particular river.

The present publication is thus a refinement of earlier efforts and in its turn it may well be found wanting and in need of further amplification, if not correction. The problem with publishing is that your mistakes are not brought to light until your efforts are made public.

Scoop Wheel, Limpenhoe Mill

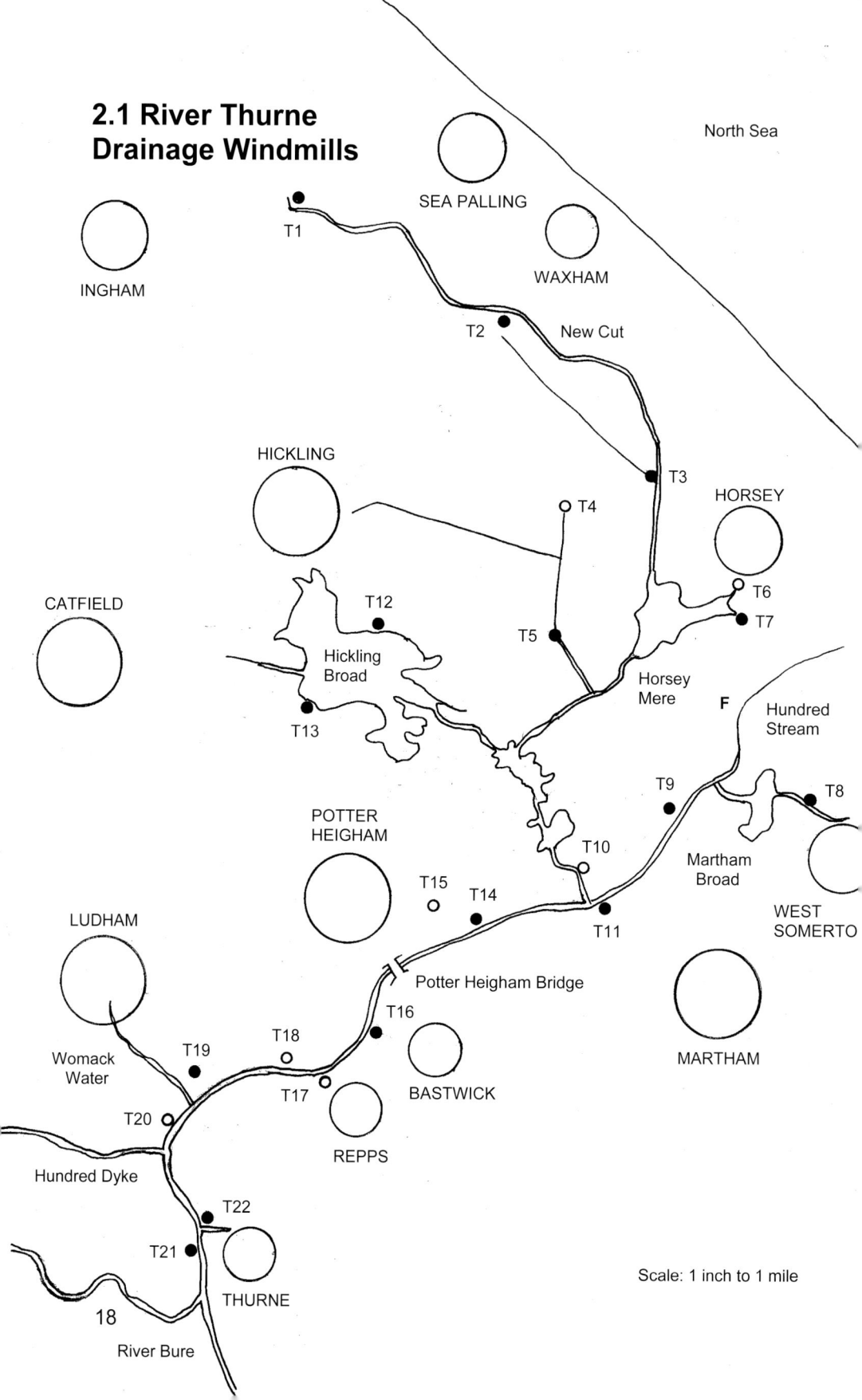

Randall's Mill, Ingham
TG 410264   T1
River Thurne, Calthorpe Marshes

1797 Faden: 'Drain Mill'

1838 OS 1" sheet 38: 'Marsh Mill'

n.d. Ingham Tithe Map: X 1843 Apportionment: (10) 'Land & Buildings', owned by 'Commissioners of Drainage'; (11) adjoining 'Mill Pightle'

1884 OS 6" sheet 41 NE: 'Marsh Mill (Draining Pump)'

Wailes: 48
Smith: 12

This is the northernmost of the drainage mills that survive on the Norfolk Broads, situated in the parish of Ingham just over a mile inland from the North Sea. It drained an area of marsh around the head of the River Thurne.

photo: polystar 2016

Wailes attributes this drainage mill to the millwright Rust, whilst Smith mentions a 'small modern pump house behind' and speaks of a 'preserved truncated red brick tower' with a 'conical corrugated sheeting roof & finial', all as shown above.

About a third of a mile nearer the sea along the road to Sea Palling, there is a mill tower at Barton Farm (TG 416267), as shown on the right. Although quite near Hempstead Marshes to the north, this mill stands on higher ground and milled corn. The ivy is in need of attention, before damage is done.

photo: polystar 2016

Lambridge Mill, Waxham
TG 432252   T2
River Thurne, Brograve Level

1838 OS 1" sheet 38:
'Lambridge Mill' X

1840 Waxham Tithe Map: X
1840 Apportionment: (204)
'Mill Yard Drain and Drift'

1884 OS 6" sheet 41 NE:
'Lambridge Mill (Draining P)'

Listed grade II 3/40

Wailes: 81
Smith: 57

About a mile south of the village of Sea Palling, at the head of the River Thurne, a widening area of marsh is now drained into New Cut along the north side of the wet area. Here Lambridge Mill was where the hard work was done keeping the area dry.

photo: polystar 2016

Grade II listed on account of its surviving internal gear, it has a four storey brick tower with an internal scoop wheel, now lost and with the pit filled in. Smith describes it as having some surviving sails. Modern OS maps show it as 'Lambrigg' Mill, as does Wailes attributing it to millwrights Rust or England.

Across the dyke to the south the 1884 6" OS map shows another 'Draining Pump' at approximately TG 432251. Its rectilinear plan on that map is confirmed by a surviving square built brick engine shed, that would have housed an auxiliary pump.

photo: polystar 2016

Brograve Mill, Waxham
TG 448236   T3
River Thurne, Brograve Level

1797 Faden: 'Waxham Drain Mill'

1837 OS 1" sheet 47: 'Mill' X

1840 Waxham Tithe Map: X
1840 Apportionment: —

1884 OS 6" sheet 41 SE: 'Brograve Mill (Draining Pump)'

Listed grade II 6/49

Wailes: 80
Smith: 58

photo: evelyn simak 2009

At the south-eastern end of Brograve Level, another brick tower mill drained these marshes into the adjoining Waxham New Cut. Known as Brograve Mill and grade II listed as of early 19th Century date, Wailes dates it to 1771, consistent with Faden.

Its derelict shell has no cap but retains parts of the patent sails and fantail and has within an octagonal windshaft and other machinery that drove an internal turbine.

Wailes gives an incorrect map reference, whilst Smith describes its very poor condition with a leaning and distorted tower, at one time tarred.

Eastfield Mill, Hickling
(TG 439233)   T4
River Thurne, Eastfield Marshes

About half a mile west of Brograve Mill, another tower mill drained marshes west of the Brograve Level, near Eastfield Farm. The 1837 1" OS map shows it as 'Eastfield Mill' with a symbol X, whilst the 1842 Hickling Tithe Map shows a circular building plan, approached by 'Mill Dike', as given in the 1843 Apportionment.
On the 1884 OS map it is shown as 'Eastfield Mill (Draining Pump)' and it appears in both Wailes (27) and in Smith's 'Some Old Sites'.

Stubb Mill, Hickling
TG 437220   T5
River Thurne, Cotton's Marsh

1837 OS 1" sheet 47:
'Stubs Mill' X

1842 Hickling Tithe Map: 'Stub Lane' approaches site
1843 Apportionment: (626) 'Mill Acres', (629) 'Little Mill Pightle' and (630) 'Great Mill Pightle' adjoin site

1884 OS 6" sheet 41 SE: 'Stubs Mill (Draining Pump)'

Listed grade II 6/7

Wailes: 28
Smith: 28

Half a mile west of Horsey Mere that collects the waters from Waxham New Cut, lies Stubb Mill. Named after the hamlet of Stubb just east of Hickling village, this brick built drainage mill was originally of three storeys.

photo: polystar 1990 N19

Since its early 19th Century origin it has been heightened, or 'hained' as Wailes puts it in the local dialect, the fourth storey being cylindrical on top of the otherwise tapering tower.

Wailes gives the original millwright as Rust and within, below a new cap fitted in 2010 (replacing the corrugated iron roof shown here) it retains the remains of internal domestic partitions along with some internal gear that originally drove an external scoop wheel.

Draining Pumps, Horsey Mere
(TG 450218 & 450219)
River Thurne, Horsey Marsh

The waters of Waxham New Cut drain to the River Thurne via Horsey Mere, where the 1884 6" OS map shows on the southern edge 'Draining Pump' in two locations near Mere Farm.

Both appear as rectangular buildings, so they are most likely to be engine sheds of some description rather than wind driven mills.

Horsey Mill, Horsey
TG 458221　　　T7
River Thurne, Horsey Marsh

1837 OS 1" sheet 47: 'Mill' X

n.d. Horsey Tithe Map: X
1840 Apportionment: —

1884 OS 6" sheet 42 SW: 'Draining Pump'

Listed grade II* 7/21

Wailes: 32
Smith: 32

photo: polystar 1990 N16

Near the village of Horsey, east of Horsey Mere and just over a mile from the North Sea coast, this mill drained an area of marshland south of the village into the Mere. It survives relatively intact and in good condition and is accordingly listed at the higher grade of II*.

Of early 19th Century origin, it was rebuilt twice, once in 1897 and later in 1912 by Dan England. Restored in 1961 by Norfolk County Council and the Society for the Protection of Ancient Buildings, it is now a National Trust property.

It had an external scoop wheel, later driven by an engine in an outbuilding, but otherwise retains the full workings and boat shaped cap missing from many other mills.

Horsey North Mill, Horsey
(TG 457225)　　　T6
River Thurne, Horsey Marsh

A little to the north of the other Horsey Mill, there was originally another one nearer the village.

Shown on the early 1" OS map of 1837 as 'Mill' with a symbol X, it also appeared on the Horsey Tithe Map of c.1840 with a symbol X, but is missing from the later 1884 6" OS map, presumably redundant by then.

West Somerton Mill, West Somerton
TG 464202   T8
River Thurne, Martham Broad

1837 OS 1" sheet 47: —

n.d. West Somerton Tithe Map: — ;
1841 Apportionment: —

1884 OS 6" sheet 54 NW: —

Listed grade II 1/46

Wailes: 66
Smith: 60

Built in 1900 by Dan England, this tower mill draining an area east of Martham Broad appears to have moved away slightly from its original position up a short dyke to the east, and is thus a definitive rebuild. The modern windows shown here inserted into the cap have now been removed.

photo: polystar 1990 N15

Otherwise with its three storey tapering brick tower with a timber boarded cap, it is fairly typical of these Norfolk drainage mills. It retains some internal machinery and originally drove a turbine pump, as one might expect from its later construction date.

North-west of this mill and south of Horsey, in 1797 Faden shows 'Old Drain M.' alongside the Hundred Stream at TG 456212, a site not shown at all on later maps.

Old Mill, West Somerton
(TG 466202)   T8a
River Thurne, Martham Broad

A little to the east of the present mill up a short dyke, the original site of West Somerton Mill was shown on the 1837 1" OS map as 'Mill' with a symbol X.
The undated West Somerton Tithe Map shows a similar symbol there, the 1841 Apportionment (68) listing 'Mill &c.' owned by 'Drainage, Commissioners of', whilst the 1884 6" OS map shows 'Draining Pump'.

Heigham Holmes Mill, Potter Heigham
TG 450202   T9
River Thurne, Eelfleet Dyke

1837 OS 1" sheet 47: 'Mill' X

1840 Potter Heigham Tithe Map: circular building plan
1841 Apportionment: —

1884 OS 6" sheet 53 NE: 'Draining Pump'

Listed grade II* 4/99

Wailes: 26
Smith: 46

photo: polystar 1990 N14

Draining an area of marsh along the north bank of the River Thurne between Eelfleet Dyke and Candle Dyke, this mill was positioned at the eastern end next to the former waterway, connecting Horsey Mere south to the river.

Although the sails are nearly gone, sufficient of the inner workings remain to justify listing this one at the higher grade of II*. Some of this gear is cast iron and includes an internal turbine pump.

Wailes gives the millwright as Smithdale, but mentions a chain guide and turbine by D. England, which must have been a later addition nearer the turn of the 20th Century, maybe when the turbine was installed.

Candle Dyke Mill, Potter Heigham
(TG 457225)   T10
River Thurne, Candle Dyke

At the western end of these marshes along the north bank of the River Thurne, there was a second drainage mill adjoining Candle Dyke, that brought the waters of Hickling Broad south to join the river.
Not shown on the 1840 Potter Heigham Tithe Map, it did appear on the 1884 6" OS map as 'Draining Pumps', adjoining both a circular and a rectangular plan.

Bracey's Mill, Martham
TG 442192   T11
River Thurne, Martham Level

1837 OS 1" sheet 47: —

1842 Martham Tithe Map: —

1884 OS 6" sheet 53 NE: —

Listed grade II 4/19

Wailes: 45
Smith: 41

photo: polystar 1990 N13

Originally with a ten blade fantail, this mill was purpose built by Dan England in 1908 to drive a wind-powered turbine pump   It is testament to other types of power not having superseded wind power by this late date, otherwise it would have been an 'Engine Shed'.

It augmented or replaced two earlier mills draining Martham Level, upstream and downstream respectively, that were both shown on the 1887 25" OS map as 'Draining Pump'.

Now converted for residential use with additional windows and an inserted water tank within the cap, there has been some inevitable loss of historic fabric, but the tower and cap retain enough to allow its listing at grade II.

Martham Pits Pump, Martham
(TG 446195)
River Thurne, Martham Level

It is thought this upstream and easternmost of a pair of replaced sites at Martham was a later auxiliary pump. Although on the 1887 25" OS map as 'Draining Pump', it had not appeared on the 1837 1" OS sheet, the 1842 Martham Tithe Map nor the 1884 6" OS.

Both are now long gone and the sites of boatyards.

Hickling Broad Mill, Hickling
TG 419221   T12
River Thurne, Hickling Broad

1838 OS 1" sheet 46: —

1842 Hickling Tithe Map: —

1884 OS 6" sheet 41 SE: 'Draining Pump'

Wailes: —
Smith: 29

photo: evelyn simak 2009

Hickling Broad is one of the wider expanses of water in the Broads area, the course of the river flowing through its middle. Peat was presumably dug from pits on both sides of the original river course, the intervening banks now long eroded away. This mill was on the north bank of the broad and not shown on the historic maps until 1884.

It is thus of mid 19th Century date and seems to have somehow escaped Wailes' notice.

Also known as Chapman's Mill or Roland Green's Mill, it is not listed since the only surviving historic remains are the bricks of the tower, used as an artist's studio and now topped by an octagonal lighthouse-like glazed turret. It had a pump outside the tower.

Original Mill, Martham
(TG 439192)   T11a
River Thurne, Martham Level

This downstream and westernmost of the pair of replaced sites at Martham was clearly the earlier and wind-powered.

On the 1884 6" OS map as 'Draining Pump', it had also appeared on the 1837 1" OS sheet as 'Mill' X, and on the 1842 Martham Tithe Map (483), mentioned in the 1844 Apportionment as 'Mill & Yard' owned by 'Commissioners of Drainage'.

Swim Coots Mill, Catfield
TG 411212   T13
River Thurne, Hickling Broad

1837 OS 1" sheet 46: X

1840 Catfield Tithe Map: circular building plan
1843 Apportionment: —

1884 OS 6" sheet 41 SE: 'Draining Pump'

Listed grade II 4/28

Wailes: 29
Smith: 14

photo: evelyn simak 2007

Across the water from the Hickling Broad Mill, this mill, also known as Ling's Mill, drained an area in the east of Catfield parish on the south bank of Hickling Broad. Like its northern counterpart, all that remains of any historic interest is a brick tower, but unlike it, this one is afforded the protection of being listed grade II.

Comprising just two storeys, it is quite squat in comparison to most with a well defined batter. It has a cement rendered section at the top of the tower and was recapped in 2007.

Water runs right through the tower which contains an internal scoop wheel that has lost its paddles. Originally it had a rare example of a seven bladed fantail.

Hickling Heath Pump, Hickling
(TG 408221)
River Thurne, Hickling Broad

At the western end of Hickling Broad there was a further site that drained the area immediately south of the hamlet of Hickling Heath for a while.

Like Hickling Broad Mill, it is not on the early maps, nor the 1884 6" OS, but only appears on the 1887 25" OS map as 'Draining Pump', so is more likely an auxiliary pump.

High's Mill, Potter Heigham
TG 430190   T14
River Thurne, Heigham Marsh

1838 OS 1" sheet 46: X

1840 Potter Heigham Tithe Map: circular building plan; 1841 Apportionment: (421) 'Mill Hill' owned by 'Drainage, Commissioners of', nearby (420) 'Mill Marsh'

1884 OS 6" sheet 53 NE: 'Draining Pump'

Listed grade II 9/100

Wailes: 49
Smith: 45

photo: polystar 1990 N21

Downstream from Candle Dyke, this mill drained an area known as Heigham Marsh on the north bank of the River Thurne, just south-east of the village of Potter Heigham itself. It now serves as a domestic dwelling.

Whilst most of the external shell survives the conversion, including a boat-shaped cap, the brickwork at high level appears different, suggesting the tower has perhaps been heightened at some point.

Wailes attributes it to the millwright Rust and also indicates that it drove a turbine pump, this last perhaps introduced at the time of heightening. Just east of here at TG 431191 the 1884 6" map shows another 'Draining Pump'.

Middle Wall Mill, Potter Heigham
(TG 425192)   T15
River Thurne, Heigham Marsh

A third of a mile along the dyke from High's Mill back towards Potter Heigham, there was in earlier times another drainage mill near Middle Wall.

This appears on the Potter Heigham Tithe Map of 1840 as a small circular building with semi-circular drainage ditch, given in the 1841 Apportionment as (365) 'Mill & Yard'.

Repps Mill, Repps with Bastwick
TG 418179   T16
River Thurne, Bastwick Marsh

1838 OS 1" sheet 46: X

n.d. Repps with Bastwick Tithe Map: circular building plan; 1840 Apportionment: 'Wall & Road' owned by 'Drainage, Commissioners of'

1884 OS 6" sheet 53 NE: 'Draining Pump'

Wailes: 57
Smith: 51

photo: polystar 1990 N22

Already converted to a house with this strange cantilevered balcony room in 1971 when Smith photographed it, this was one of the less photogenic of mills, but fortunately the cantilever section has now been removed.

It showed in relatively poor light the compromises that have to be made and damage that can ensue when a conversion is undertaken.

The survival of the tower intact with its cylindrical top section does however demonstrate that this drainage mill was probably heightened at some time. Wailes (57) states it had a scoop wheel, but has nothing further to say.

Repps Staithe Mill, Repps with Bastwick
(TG 413174)   T17
River Thurne, Repps Marsh

About half a mile downstream from Repps Mill, there was another at Repps Staithe, which appeared as 'Draining Pump' on the 1884 6" OS map.

Wailes (58) lists it as Repps Level Mill involving millwright D. England, whilst Smith includes a tower mill here in his list of 'Some Old Sites', describing it as Pug Street Mill.

Womack Water Mill, Ludham
TG 400175   T19
River Thurne, Horse Fen

1838 OS 1" sheet 46: X

1840 Ludham Tithe Map: X
1842 Apportionment: —

1884 OS 6" sheet 53 NW: 'Draining Pump'

Wailes: 44
Smith: 38

About a mile below Repps, Horse Fen on the north bank of the River Thurne was drained by a mill set some 200 yards away from the river bank. A similar distance downstream a tributary joining the Thurne from the village of Ludham to the north provides the mill's name.

By a process of elimination it is thought Wailes listed this one as Ludham Street but without a map reference, (44) in his list, but he indicates it had two scoop wheels (see T20).

Now just a derelict shell, the circular brick tower conforms to the picture we are building up of the mills hereabouts. Smith describes it as leaning over slightly, but still containing some gear inside (see p.12).

photo: arthur c smith 2005

Horse Fen Mill, Ludham
(TG 409176)   T18
River Thurne, Horse Fen

Shown with a mill symbol X on the 1838 1" OS map and as 'Draining Pump' on the 1884 6" OS map, this mill had also appeared as a circular building on the 1840 Ludham Tithe Map (214).

Wailes listed it (43) at Ludham Staithe, as did Smith in his 'Some Old Sites' list. Wailes also lists a further site (50) as Horse Fen Mill, Potter Heigham by D. England slightly further east at TG 412176, for which no other evidence is found.

St. Benet's Level Mill, Horning
TG 400156   T21
River Thurne, St Benet's Level

1797 Faden: 'Drain Mill'

1838 OS 1" sheet 46: X

1840 Horning Tithe Map: circular site only; 1841 Apportionment: (1) 'Long Mill Marsh' adjoining

1885 OS 6" sheet 53 SW: 'Draining Pump' here and at TG 400157 (square in plan)

Listed grade II* 8/45

Wailes: 65
Smith: 56

Owned by Norwich Union and restored in 1976, this four storey mill dates back to the late 18th Century and drained the east end of Horning Marsh, between where the rivers Ant and Thurne now join the Bure from the north.

photo: polystar 1990 N23

Larger than many we have seen, it has the full complement of boarded boat-shaped cap, fantail and sails, along with an external turbine pump. Wailes attributes D. England as millwright, but this probably relates to a later time when it was heightened and had a turbine fitted. The restoration will have involved the authentic reinstatement of many lost or decayed parts, but leaves us with something worthy of the higher grade II* listing.

Coldharbour Mill, Ludham
(TG 397169)   T20
River Thurne, Cold Harbour Marsh

A short section of the north-west bank of the River Thurne, between Womack Water and Hundred Dyke, was drained by a tower mill near Cold Harbour Farm. The site appeared on the 1840 Ludham Tithe Map with a windmill symbol X. Wailes (84) describes it as 'Coldharbour Mill at Womach Water' (sic) and by D. England. Excavations in 2005 revealed two scoop wheels.

Thurne Dyke Mill, Thurne
TG 401159   T22
River Thurne, Thurne Marsh

1838 OS 1" sheet 46: <u>X</u>

1843 Thurne Tithe Map: circular building plan within semi-circular ditch;
1843 Apportionment: adjoining (27) 'Mill Pightle'

1885 OS 6" sheet 53 SW: 'Draining Pump' here and another at TG 403159 further along dyke

Listed grade II* 4/62

Wailes: 73
Smith: 65

photo: polystar 1990 N11

This fine restored mill, unusually white-painted, drained the marshes on the south-east bank of the River Thurne below Repps, half a mile before it joined the Bure at Thurne Mouth.

Unlike the St Benet's Level Mill nearly opposite, this one is open to the public. Wailes (73) lists it as Morse's Mill at Thurne Staithe, attributing it to millwright D. England and with a turbine pump.

These facts probably relate to a later incarnation than that implied by the early map evidence, most likely the time it was heightened with an extra storey in the late 19th Century.

River Ant Drainage Mills

This page concludes the section covering the drainage mills of the River Thurne. We will now head back northwards again and work our way down the River Ant, until it joins the River Bure just a couple of miles west of Thurne village.
After that we will turn our attention to the River Bure itself, running from beyond Wroxham in the west all the way eastwards to join the Yare and Waveney just behind Great Yarmouth.

2.2 River Ant Drainage Windmills

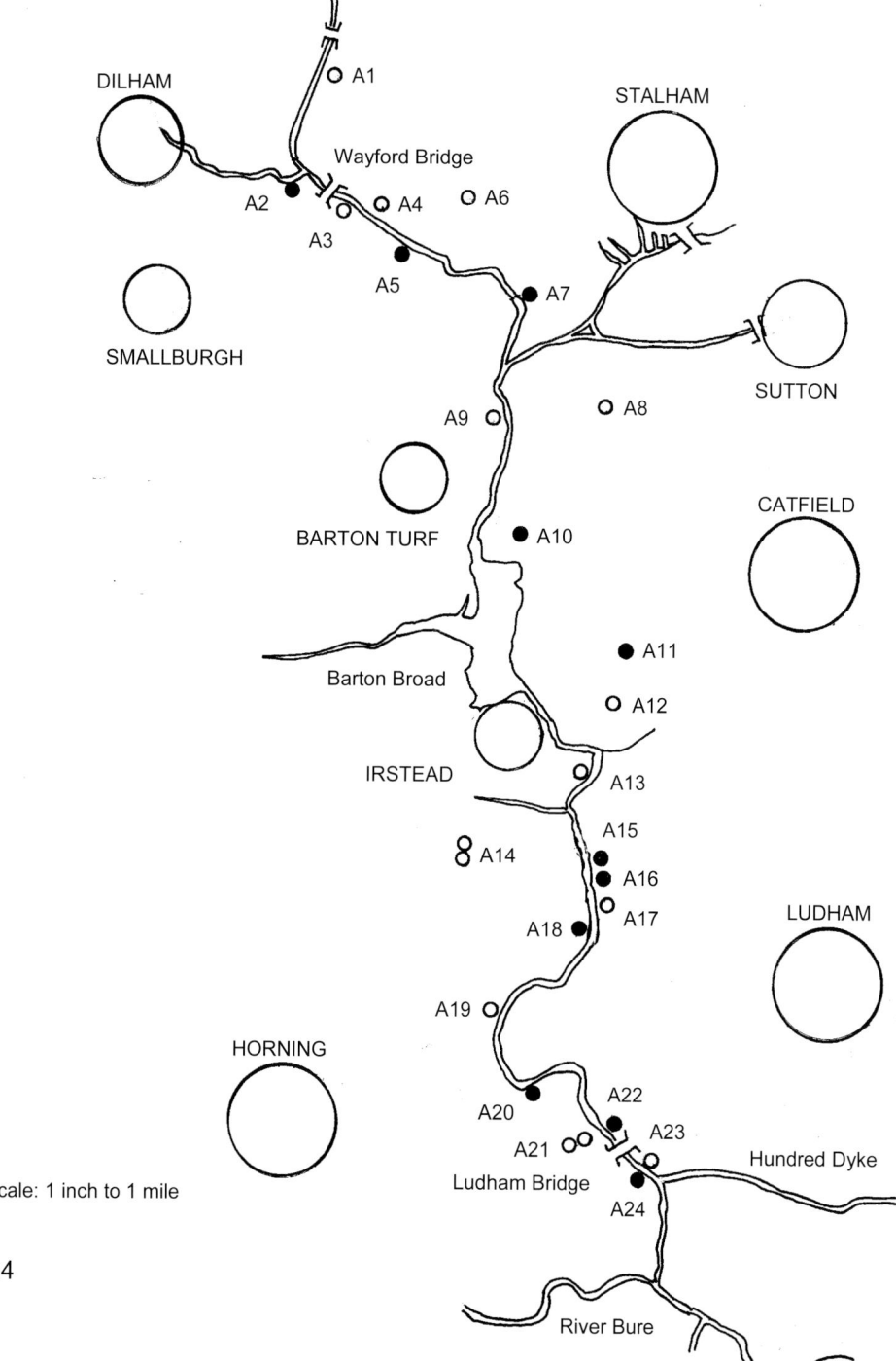

Dilham Dyke Mill, Smallburgh
TG 344248   A2
River Ant, Broad Fen

1838 OS 1" sheet 38: —

n.d. Smallburgh Tithe Map: —

1885 OS 6" sheet 40 NE: 'Draining Pump'

Wailes: 16/83
Smith: 59

Not listed, Smith (1977) notes that this mill had parts of three floors remaining and new window frames, as if it was then undergoing a conversion.

In 2016 this was at last approaching completion as the empty four storey brick tower had a new boat-shaped cap, sails and fantail ready for reinstatement. This was fitted early in 2017.

Wailes appears to have listed this mill twice: firstly under Dilham (16) with a slightly wrong map reference he attributes the millwright as W.T. England. Later at (83) he describes a mill with the right map reference as up a dyke, a quarter mile above Wayford Bridge, but attributed to D. England.

photo: polystar 2016

North Mill, Dilham
(TG 348258)   A1
River Ant, South Fen

Also north of Wayford Bridge, Wailes (94) describes a trestle mill half a mile above Wayford Bridge, up a dyke to the east. Although on the 1840 Dilham Tithe Map next to (322) 'Mill Meadow', it is shown as 'Draining Pump' on the 1886 25" OS map, but not on the earlier 1" map.
Here the River Ant becomes the North Walsham & Dilham Canal and south of it just above Honing Bridge (TG 325269), a further circular plan 'Draining Pump' is shown on the 1885 6" OS map.

Moy's Mill, Smallburgh
TG 353243   A5
River Ant, Common Fen

1838 OS 1" sheet 38: —

n.d. Smallburgh Tithe Map: —

1885 OS 6" sheet 40 NE: 'Draining Pump'

Wailes: 82
Smith: —

photo: evelyn simak 2008

Wailes' description (82) of a drainage mill half a mile below Wayford Bridge and up a dyke to the right, fits the location of this mill shown on modern OS maps, but is unfortunately given the map reference for nearby Marsh Mill at Stalham (A6).

He lists it as having a scoop wheel and attributes it to millwright Rust, but little now remains.

Nearer Wayford Bridge on the same western side of the River Ant (A3), Wailes also lists (93) the site of a trestle mill with a scoop wheel about 100 yards below the bridge (TG 349247), now the site of a boatyard. This also appeared as 'Draining Pump' on the 1885 6" OS map, but was not shown on the earlier maps.

Hollow Post Mill, Stalham
(TG 351247)   A4
River Ant, Stalham Marshes

About halfway between Moy's Mill and Wayford Bridge, but on the eastern side of the Ant, there was a mill that drained the western part of Stalham Marshes.

Given as 'Draining Pump' on the 1885 6" OS map, Wailes (98) describes it as a hollow post mill with a plunger pump, 200 yards below Wayford Bridge, up a cut to the left.

Hunsett Mill, Stalham
TG 364239   A7
River Ant, Stalham Marshes

1838 OS 1" sheet 46: X

1841 Stalham Tithe Map: circular building plan; 1846 Apportionment: (427) 'Mill and Wall'

1884 OS 6" sheet 41 NW: 'Draining Pump'

Listed grade II 5/55

Wailes: 34
Smith: 61

photo: polystar 1990 N5

Since its restoration in the mid 20th Century, this picturesque drainage mill's boat-shaped cap and sails have been fixed facing the river. Generally in good condition with a fantail and its gear inside, it now adorns a private garden.

It drained an area of marsh between the River Ant and a waterway leading to the villages of Stalham and Sutton to the east.

Dating back to the mid 19th Century, it originally had two external scoop wheels, the culverts for which remain, but are now filled in. Although earlier maps show it at this site, Wailes dates it to 1860 and calls it Durrel's Mill, but with a slightly erroneous map reference.

Marsh Mill, Stalham
(TG 359248)   A6
River Ant, Stalham Marshes

About half a mile north-west of Hunsett Mill, the 1841 Stalham Tithe Map indicates another mill draining Stalham Marshes east of the River Ant. Smith includes it in his 'Some Old Sites' list. Shown in the 1846 Apportionment (411) as 'Marsh Mill Yard', it had also appeared on the 1838 1" OS map as 'Marsh Mill' with the symbol X, and again later in 1884 as 'Draining Pump'.

Four Sites near Stalham & Sutton

River Ant, Various Marshes

Pump, Stalham
(TG 371241)
River Ant, Stalham Broad

A body of water joins the River Ant from the east just south of Hunsett Mill, serving staithes in the villages of Stalham and Sutton.

About half a mile east of Hunsett Mill, on the branch of water that forks up to Stalham village there was a 'Draining Pump' on the south bank, shown on the 1884 6" OS map sheet 41 NW. This was most likely engine driven, but might have been a small trestle mill.

Stalham Green Pump, Stalham
(TG 376244)
River Ant, Stalham Broad

The same short waterway towards Stalham Green is terminated by the modern A149 main road that runs along the route of a former railway line.

Here at its eastern end, another 'Draining Pump' was also shown on the south bank on the 1884 6" OS map sheet 41 NW. Now the site of a boat yard, this was probably also an engine driven pump of which no trace remains.

Middle Marsh Mill, Sutton
(TG 370231)   A8
River Ant, Middle Marsh

South of the waterway off the River Ant towards Stalham and Sutton a large area of marsh shows on the modern OS maps as Sutton High Fen, Big Bog, Little Bog and Middle Marsh.

The last of these has 'Old Draining Pump' shown on the 1885 6" OS map sheet 41 SW near Longmoor Farm. This is most likely the site without map reference that Wailes (92) gives for a trestle mill with scoop wheel at Sutton Marsh.

Hand Marsh Pump, Sutton
(TG 380233)
River Ant, Hand Marsh

The southern fork of the short waterway branching off the River Ant leads to Sutton village by way of Sutton Broad, and is now just a narrow channel.

On its southern side lies Hand Marsh where yet another 'Draining Pump' was shown on the 1885 6" OS map sheet 41 SW with a rectangular building. Probably just an engine house, the site is still used as such appearing on modern maps as 'Pp Ho'.

Barton Broad Mill, Catfield
TG 364220   A10
River Ant, Wood Marsh

1838 OS 1" sheet 46: X

1840 Catfield Tithe Map: —
1843 Apportionment: —

1885 OS 6" sheet 41 SW: 'Draining Pump'

Wailes: 86
Smith: 13

photo: polystar 2016

Although the well-defined stump of a tower remains, Wailes describes this drainage mill as a smock type with a map reference slightly further east. This was based on the evidence of a painting and maybe the original mill here was such.

There is some early map evidence for this mill, but Smith found here the base of a brick tower mill which he describes as one storey with a flat roof and crumbling brickwork much as it is today, presumably not enough remaining to warrant its listing.

This is one of the mills in Catfield that drained the western basin of the River Ant, as opposed to the eastern basin of the River Thurne (Swim Coots).

Barton Turf Mill, Barton Turf
(TG 361230)   A9
River Ant, Barton Marsh

About half a mile north of Barton Broad, the 1839 Barton Turf Tithe Map shows a mill draining marshes on the western bank of the River Ant, up a short dyke.

Shown in the 1840 Apportionment (60) as 'Mill and Wall', it had also appeared on the 1838 1" OS map with the symbol X, and again later in 1885 as 'Draining Pump'.

Middle Marsh Mill, Catfield
TG 372211   A11
River Ant, Catfield Marshes

1838 OS 1" sheet 46: X

1840 Catfield Tithe Map: —
1843 Apportionment: —

1885 OS 6" sheet 41SW: 'P'

1886 OS 25":
'Draining Pump'

Wailes: —
Smith: —

photo: polystar 2016

One of a pair of mills on the mile wide expanse of Catfield Marshes, this one is over half a mile away from the River Ant and drained Middle Marsh into a network of dykes connecting to the river further west. It seems somehow to have been overlooked by both Wailes and Smith.

Not on the modern OS 1.25" Landranger map, it is shown on the modern OS 2.5" Explorer map as 'Middle Marsh Drainage Mill' with the usual symbol X, but not quite in the right place.

Now just a derelict red brick tower, it has lost the brickwork above the original doorway, but remains roofed in. The top brickwork courses have been repaired and capped with a slim modern lead flashing.

Little Fen Mill, Catfield
(TG 371207)   A12
River Ant, Catfield Marshes

Also shown with a mill symbol X on the 1838 1" OS map, but not on the Catfield Tithe Map, and then as 'Draining Pump' and 'P' on the 25" and 6" maps respectively, this mill had a similar history to Middle Marsh Mill, but has now gone.

It was nearer the river and drained the southern part of Catfield Marshes.

Clayrack Mill, Ludham
TG 369193 (TG 367149)   A15
River Ant, How Hill Marsh

1838 OS 1" sheet 46: —

1840 Ludham Tithe Map: —

1884 OS 6" sheet 53 NW: —

1885 OS 6" sheet 53 SW: 'Draining Pump' (at former Ranworth site)

Wailes: —
Smith: —

photo: polystar 1990 N8

Obviously not on any of the old maps in its current position, this hollow post type drainage mill was moved to its present site just north of Boardman's Mill during the 20th Century.

It originally drained Ranworth Marshes south of the River Bure, to the east of Malthouse Broad and was shown at that site on the 1885 25" OS map.

Unfortunately the materials used at its 'restoration' were a little substandard and although captured on film in 1990, it is now reputedly in less good condition.

Hall Fen Mill, Irstead
(TG 369201)   A13
River Ant, Hall Fen

About half a mile north of Clayrack Mill, the 1838 1" OS map shows a mill symbol X, draining Hall Fen on the western bank of the River Ant just below Irstead village. This mill also appears on the 1884 6" map as 'Draining Pump'.

Two further 'Draining Pump's (A14) are shown at Irstead Street (TG 359195). Wailes (96/97) indicates two hollow post mills there at TG 358196, one with a plunger pump, the other with a scoop wheel and built in 1920.

Boardman's Mill, Ludham
TG 369192   A16
River Ant, How Hill Marsh

1838 OS 1" sheet 46: —

1840 Ludham Tithe Map: —

1884 OS 6" sheet 53 NW:
'Draining Pump'

1887 OS 25":
'Draining Pump (Wind)'

Listed grade II* 8/66

Wailes: 90
Smith: 37

photo: polystar 1990 N9

This grade II* drainage mill is a fine example of the trestle type of mill dating from the late 19th Century. Also known as Skeleton Mill on account of its open frame, its miniature cap, sails and fantail are all the result of modern restoration.

Wailes attributes it to millwright D. England, whilst Smith's 1977 photograph shows it without the sails, cap or fantail, indicating a considerable degree of restoration to the top of the workings.

Now relatively intact, it has a turbine pump within a brick well at the base, driven through a crown wheel with universal joint from a cast iron vertical shaft.

How Hill Mill, Ludham
(TG 370191)   A17
River Ant, How Hill Marsh

Although not on the early maps, the 1884 6" and 1887 25" OS maps show a further 'Draining Pump' a little to the south of Boardman's Mill.

Wailes lists a mill (33) at How Hill Staithe with a similar map reference (TG 370192) and says it had two scoop wheels, but this probably refers to Turf Fen Mill a little further south (see A18).

Turf Fen Mill, Irstead
TG 369188   A18
River Ant, Turf Fen

1838 OS 1" sheet 46: —

1839 Irstead Tithe Map: —

1884 OS 6" sheet 53 NW:
'Draining Pump'

1887 OS 25":
'Draining Pump (Wind)'

Listed grade II* 8/10

Wailes: 33
Smith: 6

A short distance downstream from Boardman's Mill, on the west bank of the River Ant, this red brick tower mill has a similar story. Wailes seems to have given it the wrong map reference, whilst Smith's 1971 photo shows it with little in the way of cap and just the shafts of the sails up top.

photo: polystar 1990 N4

Now grade II* listed, with a date of 1880, the restored version boasts a complete set of sails and a rotating white boarded boat-shaped cap with six bladed fantail.

All its machinery survives including a cast iron windshaft, crown wheel and pit wheel, which drove two external scoop wheels through gearing with two different ratios. These last make it unique enough for the higher grade II* listing.

Two Pumps, Ludham
(TG 366185 & TG 370187)
River Ant, Ludham Marsh

A marshy area just south of Turf Fen Mill on the opposite (eastern) bank of the River Ant was at one time drained at two sites.

Each appeared on the 1887 25" OS map as 'Draining Pump', and both together were shown on the 1885 smaller scale 6" map marked 'Draining Pumps', but not on the earlier maps.

Neave's Mill, Horning
TG 365176   A20
River Ant, Horning Marsh

1838 OS 1" sheet 46: —

1840 Horning Tithe Map: —

1884 OS 6" sheet 53 NW: 'Draining Pumps'

Listed grade II 8/40

Wailes: 30/41
Smith: 5

photo: colin mitchell 2005

This four storey brick tower mill was built c.1870 on the right bank of the River Ant about half a mile above Ludham Bridge. It now has an aluminium-clad boat-shaped cap which protects a good deal of original internal gear.

Correctly map referenced by Wailes (30) as on Horning Marshes, he gives no further detail of this item in his list. He then repeats himself at entry (41) with a description of a tower mill with a turbine pump half a mile above Ludham Bridge.

Smith then complicates matters by including a tower mill here in his 'Some Old Sites', but illustrating this mill at his entry (5), describing it as 1¼ miles up from Ludham Bridge at Barton Turf, and with the Browns Hill Mill map reference. This might be explained by the modern OS map having a mill symbol at Browns Hill, but not here at Neave's Mill?

Browns Hill Mill, Irstead
(TG 362182)   A19
River Ant, Turf Fen

Described as 'Irstead Mill' on the 1838 1" OS map, 'Mill and Yard' in the 1839 Irstead Tithe Apportionment (4) and as 'Draining Pump (Wind)' on the 1887 25", this mill formerly drained an area just north of Horning Marshes.

Listed correctly by Wailes (42) as 1 mile above Ludham Bridge, he fails to give a map reference.

Bridge Mill, Ludham
TG 372172　　A22
River Ant, Ludham Marsh

1838 OS 1" sheet 46: —

1840 Ludham Tithe Map: —
1842 Apportionment: —

1884 OS 6" sheet 53 NW:
'Draining Pump'

1887 OS 25":
'Draining Pump (Wind)'

Wailes: 40
Smith: 39

photo: colin mitchell 2005

Just upstream of Ludham Bridge, this mill at one time drained Ludham Marsh on the eastern bank of the River Ant. Its remains are still visible from the bridge, but a little hidden now by trees. During WWII it was converted into a two storey pill-box with gun slits in lieu of windows.

Wailes (40) lists it as having a turbine pump, whilst Smith describes much as we see today: a three storey derelict red brick tower, leaning over and with some beams inside.

Presumably not listed on account of its poor condition and relative lack of surviving machinery, it is also not on the early maps suggesting a relatively later construction date. Its survival is thus perhaps under threat.

Two Mills, Horning
(TG 369171 & 370171)　A21
River Ant, Horning Marsh

An area of marsh adjoining Horning Marsh, just north of the main road over Ludham Bridge, on the western bank of the River Ant was at one time drained by two mills.

Each appeared on the 1887 25" OS map as 'Draining Pump (Wind)', and together on the 1884 6" OS as 'Draining Pumps', but not on the earlier maps.

Ludham Bridge Trestle Mill, Horning
TG 374169   A24
River Ant, Horning Marsh

1838 OS 1" sheet 46: —

1840 Horning Tithe Map: —

1885 OS 6" sheet 53 SW: —

Wailes: 91
Smith: 40

On the bank opposite the site of Bridgefen Mill there remain a few brickwork piers and an iron wheel on a spindle in a concrete pump shaft, as shown here.

These are the remains of a later trestle mill with a turbine pump, not on the maps. This was attributed by Wailes to D. England.

photo: colin mitchell 2006

Bridgefen Mill, Ludham
(TG 374170)   A23
River Ant, Ludham Marsh

About as far downstream as Bridge Mill is above Ludham Bridge, there was until its demolition in the 1960's another large drainage tower mill with a scoop wheel on this same eastern bank of the River Ant.

This appeared on the 1838 1" OS map and the 1840 Ludham Tithe Map with a symbol X, described in the 1842 Apportionment (706) as 'Drain Windmill &c.' owned by 'Drainage, Commissioners'. It appears on the 1885 OS 6" as 'Draining Pump'. It was described by Wailes (39) as by D England and included by Smith in his 'Some Old Sites' list as a tower mill.

Horning Hall Mill, Horning
(TG 376161)
River Ant, Horning Marsh

Just before the River Ant ceases to exist where it joins the Bure, there was a drainage site on the western bank near Horning Hall.

Not on the early maps and not listed by Wailes nor Smith, it appeared as 'Draining Pumps' on the 1884 6" OS map sheet 53 SW. Probably an engine, it formerly drained the southernmost corner of Horning Marsh.

2.3 Upper River Bure Drainage Windmills

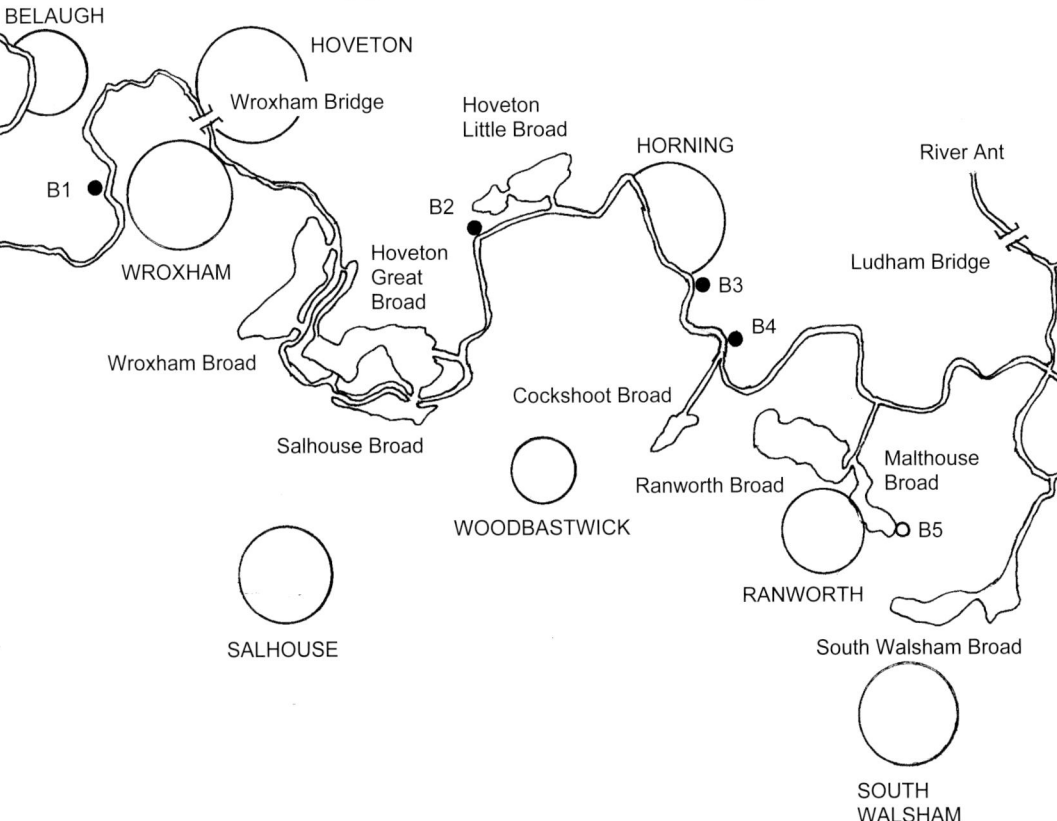

River Bure Drainage Mills

Having now completed both the Rivers Thurne and Ant down to where they join the River Bure, we must now head westwards.

We will start our survey of the River Bure beyond Wroxham and follow it on its long journey eastwards towards Great Yarmouth and the sea, taking it in three sections.

Scale: 1 inch to 1 mile

Belaugh Old Hall Mill, Belaugh
TG 293176   B1
River Bure, Belaugh Broad

1838 OS 1" sheet 46: —

1839 Belaugh Tithe Map: —

1887 OS 6" sheet 52 NW: 'Draining Pump'

Wailes: —
Smith: 7

photo: polystar 2016

More like a tree to look at than a mill, this derelict red brick tower mill at Belaugh Old Hall above Wroxham is encased in ivy, which over time will do it no good whatsoever. Photographs thus show little of interest to the mill enthusiast, which might be why Wailes seems not to have found this one.

Like the Ludham Bridge Mill (A22) this one is presumably not listed on account of its poor condition and relative lack of surviving machinery. Like that example it is also not on the early maps suggesting a relatively late construction date and its survival equally at risk.

Four More Sites above Wroxham
(TG 259208, 285188, 282172 & 282173)
River Bure

Above Wroxham, the River Bure is barely tidal and runs in a relatively narrower valley than below where we find Wroxham Broad. Small riverside areas of wetland here also needed draining and Faden's 1797 map shows a 'Drain Mill' at Great Hautbois, TG 259208. Although not shown on the early maps, three further drainage sites appeared on the 1887 6" OS map sheet 52 NW, two as 'Draining Pump', the last as 'Draining Pump (Disused)', which were most likely engine driven.

Dydall's Mill, Hoveton
TG 326171   B2
River Bure, Hoveton Marshes

1838 OS 1" sheet 46: —

1841 Hoveton Tithe Map: —
184? Apportionment: —

1885 OS 6" sheet 52 NE: 'Draining Pump'

Wailes: 31
Smith: 33

photo: david dixon 2015

Described by Wailes (31) as at Horning Street, this mill formerly drained Hoveton Marshes about a mile west of Horning village. He attributes D. England as millwright and gives a slightly erroneous map reference.

Smith fortunately gives us the back story, describing the mill as having been burnt out around 1913 and converted for residential use in 1934. The three storey red brick tower is now topped by a lighthouse like octagonal pavilion with a balcony up amongst the neighbouring trees.

Presumably there is no remaining internal gear, but the new use has certainly secured the tower's future without too much loss.

> **Two Further Sites, Wroxham**
> (TG 300181 & 309176)
> *River Bure, Bridge & Wroxham Broads*
>
> Not shown on the 1838 1" OS nor the 1839 Wroxham Tithe maps, a 'Draining Pump' was shown on the 1887 6" map sheet 52 NW at each of these sites.
>
> These were also probably engine driven and formerly drained areas near Wroxham Bridge and at the north end of Wroxham Broad (now a boat yard), both on the south bank of the River Bure.

Horning Ferry Mill, Horning
TG 345167   B3
River Bure, Church Marsh

1838 OS 1" sheet 46: —

1840 Horning Tithe Map: —

1887 OS 6" sheet 52 SE: —

Listed grade II 7/49

Wailes: 87
Smith: 31

This smock mill is not all it appears, concealing within its bell-shaped boarded tower, atop an octagonal modern base, a much more slender octagonal smock mill as illustrated by Wailes in his plate XXXIX Fig. 1.

The internal framing and brakewheel survive, but the upright shaft and turbine pump it once drove are long gone. The cap is fairly original, but the rest is in the words of the list description 'entirely false and late C20'. The list description is qualified by the words 'Included for rarity value as a smock mill', which should perhaps have been thought about before it was converted.

The Horning Tithe Map is remarkable perhaps for its lack of windmills, with only that at St. Benet's Level appearing in plan, and none at all listed in the apportionment, as if they did not exist in this parish.

photo: polystar 1990 N25

Another Site, Horning
(TG 345163)
River Bure, Church Marsh

A quarter mile to the south of Horning Ferry Mill, another draining pump once graced the northern bank of the River Bure, at a site now occupied by a boat yard.

Not on the early maps it appeared on the 1887 6" OS map sheet 52 SE as 'Draining Pump', and again was most likely engine driven.

Hobb's Mill, Horning
TG 347163   B4
River Bure, Church Marsh

1838 OS 1" sheet 46: —

1840 Horning Tithe Map: —

1887 OS 6" sheet 52 SE: —

Listed grade II* 7/39

Wailes: 89
Smith: 30

photo: colin mitchell 2004

Described by Wailes as driving a scoop wheel, this trestle mill certainly matched Smith's description as 'derelict and leaning over', when his photograph was taken in 1978. The uniqueness of a trestle mill with a scoop wheel is the reason for its higher grade II* listing.

Fortunately the list description is able to report better news as the superstructure other than stocks and sails was restored in 1983, with the scoop wheel to follow.

Its non-appearance on the early maps accords with its late 19th Century date, a time when four brick piers supporting four cross-braced canted posts was all that was needed, rather than the full brick tower.

Small Mill, Ranworth
(TG 362147)   B5
River Bure, Ranworth Marshes

Below Horning a stretch of the River Bure is now mainly wooded on the south side containing Ranworth and Malthouse Broads. A 'Draining Pump' appeared at the east end of the latter on the 1887 OS map, described by Wailes (68) as a tower mill (in South Walsham!) and included in Smith's 'Some Old Sites'.

Two other sites here (TG 353160 & 367149) also appear as 'Draining Pump', the latter now re-erected as Clayrack Mill on the River Ant (A15).

Middle River Bure Drainage Windmills

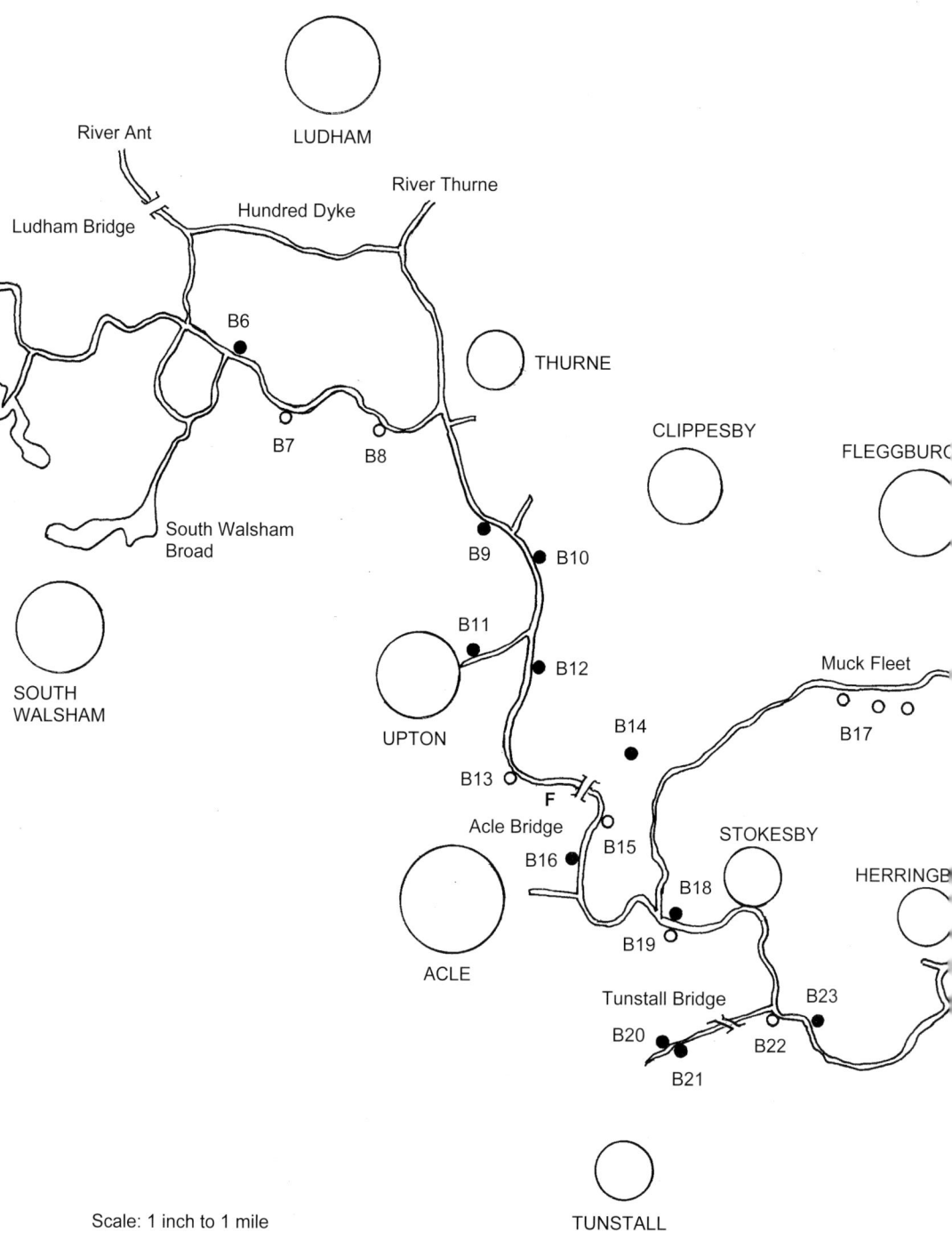

St. Benet's Abbey Mill, Horning
TG 380158   B6
River Bure, Horning Marsh

1797 Faden: 'Mill & Gate'

1838 OS 1" sheet 46: 'St. Bennet's Mill'

1840 Horning Tithe Map: —

1885 OS 6" sheet 53 SW: 'Draining Pump (Disused)'

Listed grade I 8/41

Wailes: 64
Smith: 55

photo: polystar 1990 N3

Purportedly built between 1728 and 1735, this red brick tower mill is certainly not of the 8 storeys that are attributed to it in the list description. The very high grade I status is on account of the medieval abbey gatehouse the mill partly swallows up.

An early tower mill then, apparently used to crush colza seed for the oil used in lamps, but possibly converted for drainage use at some time later.

Some 19th Century pictures appear to show a scoop wheel here, but Smith could find no evidence of a bricked up aperture where the shaft would have pierced the tower, so it remains a mystery, but one well worth a visit.

Debbage's Mill, South Walsham
(TG 385151)   B7
River Bure, South Walsham Marshes

Across the river from St Benet's, the early OS maps show a X symbol (1838) and 'Draining Pump' (1885), indicating a mill draining a very large area of marsh here.

Wailes (67) describes it as 'Debbage's Mill', a tower mill with a date c.1820, and Smith subsequently includes it in his 'Some Old Sites'. Still 'Draining Pump' on modern maps this is a newer installation.

Tall Mill, Upton
TG 405141　　　B9
River Bure, Upton Marshes

1838 OS 1" sheet 46: 'Upton Mill' X

1839 Upton Tithe Map: X
1841 Apportionment: (258) 'Mill & Yard' owned by 'Commissioners of Drainage'

1885 OS 6" sheet 53 SW: 'Draining Pump'

Listed grade II 2/79

Wailes: 79
Smith: 68

photo: polystar 1990 N26

This mill's four storey tower justifies its name and the change of batter at the top suggests a heightening. Wailes has it with a turbine and attributes it to millwright Smithdale, but gives the map reference for Fishley Mill (B13).

Smith corrects this error and notes that it had been converted into a house. Fortunately this has impinged little on what we see today with the tarred brick tower and its boat shaped cap with fantail staging still dominant.

An engine house dating from the 1840's is close by and also listed grade II (see p.11). This originally had a steam engine driving a large scoop wheel, but was later given a diesel driven turbine pump.

An Earlier Mill, Upton
(TG 395149)　　　B8
River Bure, Upton Marshes

Roughly midway between Debbage's Mill and Tall Mill, another drained Upton Marshes south of the River Bure, just upstream from Thurne Mouth, given by Wailes (78) as by W.T.England.
This was shown on the 1838 1" OS map with the usual X symbol near 'Marsh Farm' and also on the 1839 Upton Tithe Map (283), apportioned in 1841 as 'Mill Yard', but not on the later 1885 6" OS.

Wiseman's Mill, Ashby with Oby
TG 410138   B10
River Bure, Oby Marsh

1797 Faden: 'Oby Drain Mill'

1838 OS 1" sheet 46: 'Oby Mill' X

1843 Thurne, Ashby & Oby Tithe Map: circular building plan; 1843 Apportionment: (224) 'Mill & Yard'

1885 OS 6" sheet 65 NW: 'Draining Pump'

Listed grade II* 6/4

Wailes: 47
Smith: 3

This mill's higher grade II* listing is based on its early date of 1753 and the survival of much of the internal gear. With some later gearing connecting the turbine pump to an adjacent steam engine shed, this tells a more complete story than many mills.

photo: polystar 1990 N2

At one time a sawbench was also operated here and the reduced batter of the top storey suggests this is yet another mill that has been heightened, possibly when the pump was installed.

Wailes attributes this mill to Robert Martin, who operated out of Beccles. Now with the protection of a tarpaulin cap, the gear is less at risk than when this photograph was taken.

Boundary Farm Site, Ashby with Oby
(TG 403151)
River Bure, Oby Marsh

The parish boundary between Thurne and Oby passes along a small dyke with a staithe, off the River Bure just south of Thurne Mouth.

A 'Draining Pump' is shown here on the Oby side, near Boundary Farm, on the 1885 6" OS map sheet 53 SW, but not on the earlier maps, so this site was probably engine driven.

Palmer's Mill, Upton
TG 403129 (TG 404103)   B11
River Bure, Upton Marshes

1838 OS 1" sheet 46: —

1839 Upton Tithe Map: —

1885 OS 6" sheet 65 NW: —

Wailes: 95
Smith: 69

photo: colin mitchell 2006

Originally sited just south of Acle (at approx TG 404103), this hollow post mill with a plunger pump was described by Wailes (95) in its original location. He attributes it to Whitmore & Binyon with a later rebuild by Smithdale.

Relocated in the 1970's, presumably because it lay in the path of the proposed by-pass south of Acle village, it was re-erected adjoining Upton Dyke nearly two miles further north and later completely restored.

Although privately owned it is accessible both by boat and footpath, the only surviving example of its type in the Broads area.

Fishley Mill, Fishley
(TG 407118)   B13
River Bure, Fishley Marshes

About a mile south of Upton Dyke the 1838 1" OS map shows 'Fishley Mill' with a symbol X, whilst the 1841 Fishley Tithe Apportionment (19) lists 'Garden with site of Drainage Mill'. The 1885 6" OS map also shows 'Draining Pump' all indicating a mill here draining Fishley Marshes.

Wailes (79) and Smith (Some Old Sites) both mistakenly describe a mill at this location as 'Tall Mill, Upton' (see B9, p.54), in addition to their other entries for the latter.

Clippesby Mill, Ashby with Oby
TG 409128   B12
River Bure, Clippesby Marsh

1797 Faden: 'Drain Mill'

1838 OS 1" sheet 46: 'Clippesby Mill' X

1843 Thurne, Ashby & Oby Tithe Map: circular building plan; 1843 Apportionment (233) 'Mill & Yard'

1885 OS 6" sheet 65 NW: 'Draining Pump'

Listed grade II 6/5

Wailes: 14
Smith: 2

photo: polystar 1990 N1

Just south of Upton Dyke on the east bank of the River Bure, Wailes gives this four storey red brick tower mill a slightly erroneous map reference, which Smith fortunately corrects.

By 1990 the stocks Smith describes have gone, but the boat-shaped cap with attached gallery remains intact along with a good deal of internal machinery.

Although shown by Faden in 1797, the list description has this mill with a scoop wheel as of mid 19th Century date and now converted to residential use, apparently without the loss of fabric and aesthetic appeal we have seen in some conversions.

Another Site, Ashby with Oby
(TG 413131)
River Bure, Clippesby Marsh

A little north-east of Clippesby Mill, another is indicated by 'Draining Pump' on the 1884 6" OS map sheet 65 NE, away from the river bank.

This late, with no earlier reference on the old maps, it might simply be an engine site, but its location would be difficult to service with fuel supplies.

Fleggburgh Mill, Fleggburgh
TG 418119   B14
River Bure, Fleggburgh Marsh

1838 OS 1" sheet 46:
'Burgh Mill' X

1838 Fleggburgh Tithe Map:
X

1884 OS 6" sheet 65 NE:
'Draining Pumps'

Wailes: 2/17
Smith: 18

photo: evelyn simak 2009

Just up the road from Wey Bridge, also known as Acle Bridge, this former drainage mill stands near the A1064 road from Acle to Fleggburgh and originally drained an area between those two villages a quarter mile from the River Bure. Given in the plural on the 1884 6" map, with a rectangular building alongside the circular one, it obviously had an auxiliary pump.

Wailes (2) attributes this mill to millwright Smithdale, whilst Smith speaks of a tarred red brick tower with a glazed lookout beneath the boat-shaped cap. Wailes (17) also lists another Fleggburgh (also called Billockby) Mill which he attributes to millwright W.T. England and notes that the gear has been removed to Ashtree Mill (B32). Not listed on account of its conversion and lack of internal gear, it is now painted white and has a modern house attached.

Bridge Farm Mill, Fleggburgh
(TG 416114)   B15
River Bure, Fleggburgh Marsh

In 1797 Faden shows this site as 'Drain W. Mill', along with another across the river above the bridge at TG 411117.
Appearing on the 1884 6" OS map as 'Draining Pump', Wailes (69) describes Calthorpe's Mill, Stokesby with a scoop wheel here, which was also included in Smith's 'Some Old Sites'.

Hermitage Mill, Acle
TG 413110   B16
River Bure, Hermitage Marshes

1838 OS 1" sheet 46: X

1838 Acle Tithe Map: circular building plan;
1840 Apportionment (312) 'Mill Hill with Yard'

1884 OS 6" sheet 65 NE: 'Draining Pump'

Wailes: 72
Smith: 1

Adjoining Mill House Farm on the western bank of the River Bure below Acle Bridge, this former wind driven mill drained Hermitage Marshes north of Acle Dyke between the river and the village of Acle. It now houses an electric pump.

photo: evelyn simak 2013

Wailes says it had a scoop wheel and describes it as Charlie Water's Mill at Thrigby, attributing Smithdale as the millwright.

Smith describes its truncated red brick tower, now with a flat roof and double doors to facilitate its use as a farm store, not the most picturesque end to a building with such potential to enhance the environment.

Muck Fleet Mills, Fleggburgh
(TG 438125, 442124 & 445124)   B17
Muck Fleet, off River Bure

The area inland of the coastal villages of Hemsby, Scratby and Caister contains Ormesby, Rollesby and Filby Broads that drain south-westward into the River Bure by way of Muck Fleet south of Fleggburgh.

Here the 1837 1" OS map sheet 47 shows two mill symbols X adjoined by the text 'Mills'. The later 1884 6" OS map shows a third one to the east with a circular building plan.

Commission Mill, Stokesby
TG 422104   B18
River Bure, Stokesby Marsh

1797 Faden: 'Stokesby Drain Mill'

1838 OS 1" sheet 46: X

1840 Stokesby Tithe Map: circular building plan
1841 Apportionment: —

1884 OS 6" sheet 65 SE: 'Draining Pump'

Listed grade II 6/48

Wailes: 70
Smith: 62

photo: polystar 1990 N36

Just east of where Muck Fleet discharges into the River Bure, on the north bank, this red brick tower mill is now capped by a flat roof atop short brick piers, rather than by a traditional boat-shaped cap.

The tower remains intact with its original window openings and it is grade II listed, despite a conspicuous lack of internal machinery.

Of early 19th Century date, Wailes attributes it to millwright William Rust of Martham. It was converted to residential use in the 1980's.

Acle Mill, Acle
(TG 422103)   B19
River Bure, Calthorpe Level Marshes

Directly opposite Commission Mill on the south bank of the River Bure, there was another mill shown as 'Acle Drain Mill' by Faden in 1797.
This was shown on the 1838 1" OS map as 'Acle Mill' with the usual X symbol and on the 1838 Acle Tithe Map (663) as a circular building plan, apportioned in 1840 as 'Mill Hill and Mill', occupied by 'Robert England'. It also appeared on the 1884 6" OS map as 'Draining Pumps' with a rectangular building alongside the circular one.

Tunstall Dyke Mill, Tunstall
TG 422092   B20
River Bure, Calthorpe Level Marshes

1838 OS 1" sheet 46: X

1847 Tunstall Tithe Map: X
1848 Apportionment: (59) 'Mill and Yard'

1884 OS 6" sheet 65 SE: 'Draining Pump'

Listed grade II 4/65

Wailes: 1
Smith: 66

This derelict tarred red brick tower is situated on the north side of Tunstall Dyke, a nearly mile long inlet west of the River Bure, that passes under both the main A47 road and the more recent railway line on its way to a staithe near the village.

Although Smith describes it as a shell only, the list description indicates the survival of the curb and track at the top as well as the cast iron pit wheel and horizontal drive linking to the scoop wheel at the base. This is presumably why it was listed contrary to initial impressions, although its early date will have probably tipped the balance.

photo: evelyn simak 2008

Two Sites, Tunstall & Acle
(TG 421091 & 408091)
River Bure, Tunstall & Damgate Marshes

Further up Tunstall Dyke on the southern side there was another drainage site slightly nearer the village than the Tunstall Dyke Mill. Not on the early maps, it appeared as 'Draining Pump' with a rectangular plan on the 1884 6" OS map and augmented the smock mill's drainage of Tunstall Marshes.

Further west near Damgate on the parish boundary with Acle, another 'Draining Pump' was also probably engine driven.

Tunstall Dyke Smock Mill, Tunstall
TG 423092   B21
River Bure, Tunstall Marshes

1797 Faden: 'Drain W. Mill'

1838 OS 1" sheet 46: X

1847 Tunstall Tithe Map: X
1848 Apportionment: (91) 'Mill and Yard'

1884 OS 6" sheet 65 SE: circular ditch only

Listed grade II 4/66

Wailes: 85
Smith: 67

photo: evelyn simak 2008

Just across Tunstall Dyke from the remains of the tower mill, there was a further mill on the south bank. The list entry describes this one as the 'Only surviving drainage Smock Mill in Norfolk', its rarity the main reason for listing.

The early map evidence suggests that this smock mill originally drained the western part of Tunstall Marshes that lay south of the dyke. Octagonal in plan, with a trestle type frame clad in weather-boarding, it is now truncated to just two storeys and has a number of other buildings adjoining. Wailes says it drove a turbine pump.

Tunstall Bridge Mill, Tunstall
(TG 432095)   B22
River Bure, Tunstall Marshes

Another mill draining Tunstall Marshes was shown by Faden in 1797 as 'Drain Mill', where Tunstall Dyke met the River Bure adjoining Tunstall Bridge, now a modern 'Pumping Station'.

photo: evelyn simak 2011

Shown as X on both the 1838 1" OS map and the 1847 Tithe map, the 1848 apportionment (137) gives us 'Mill and Yard' owned by 'Isaac Everitt'. Wailes (75) describes it as a tower mill driving a scoop wheel, dating from 1818 and Smith includes it in his 'Some Old Sites'.

Old Hall Mill, Herringby
TG 437095   B23
River Bure, Herringby Marsh

1797 Faden: 'Drain W. Mill'

1837 OS 1" sheet 47: X

1840 Stokesby & Herringby Tithe Map: enclosure shown; 1841 Apportionment: (333) 'Mill Marsh' adjoining

1884 OS 6" sheet 65 SE: 'Draining Pumps'

Wailes: 71
Smith: 63

The next marsh on the north bank of the River Bure downstream from Stokesby was Herringby Marsh, drained at its western end by this mill half a mile south of Stokesby Hall. Next to a marshman's cottage, Wailes lists a tower mill driving a scoop wheel, but gives no further detail.

photo: pierre terre 2008

Smith's photograph from 1973 shows a little more undergrowth than was there in 1990, his description being of a small derelict tarred red brick tower.

It was shown in the plural on the 1884 OS map with both a circular and a rectangular plan, the latter presumably an engine driven auxiliary. Definitely in existence by the early 19th Century, it is not listed on account of its poor condition and lack of internal gear.

Herringby Hall Pump, Herringby
(TG 447101)
River Bure, Herringby Marsh

About a mile downstream from Old Hall Mill, the eastern end of Herringby Marsh was drained by another auxiliary pump a short distance away from the river bank near Herringby Hall.

Not on the early maps, it was shown as 'Draining Pump' on the 1884 6" OS map sheet 65 SE, so was probably built to augment the western one.

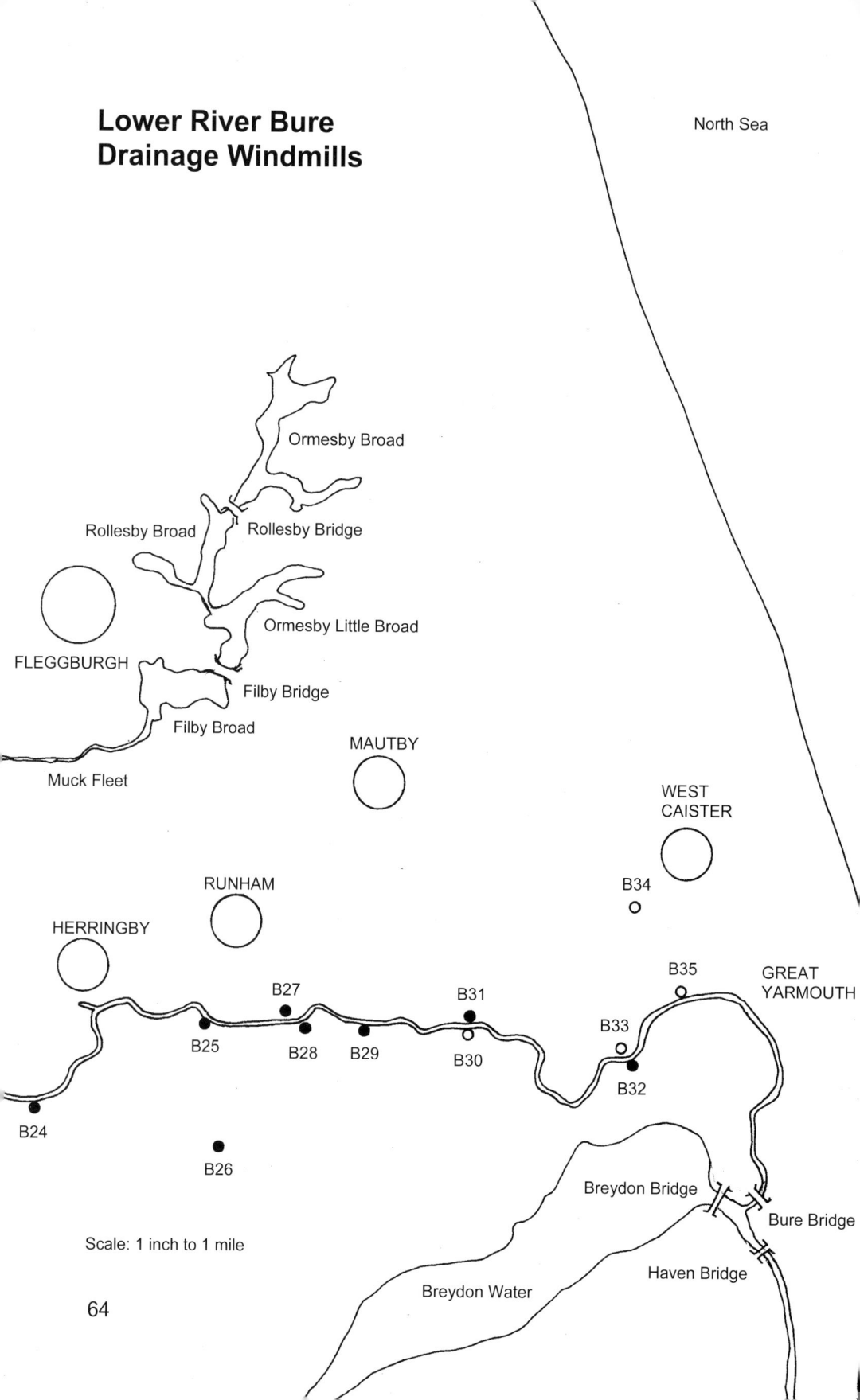

Stracey Arms Mill, Tunstall
TG 442090   B24
River Bure, Tunstall Marshes

1797 Faden: 'Drain W. Mill'

1837 OS 1" sheet 47: X

1847 Tunstall Tithe Map: X
1848 Apportionment: (149) 'Mill Drain and Yard' owned by 'Sir Edward Stracey'

1884 OS 6" sheet 65 SE: 'Draining Pump'

Listed grade II* 4/48

Wailes: 76
Smith: 64

In between the two Herringby sites, on the opposite southern bank of the River Bure, this four storey mill drained the eastern part of Tunstall Marshes, very near where the 'New Road' Acle turnpike cut across the marshes in 1830.

photo: polystar 1990 N27

Wailes gives an alternative name of Arnup's Mill and dates it to 1883, attributing it to millwright Barnes, who most likely installed the turbine pump at that time. This was probably a late Victorian rebuild, as the early map evidence suggests a longer timescale.

Smith notes the more recent 1961 restoration and the general good condition internally, the reason for the higher grade II* listing. However, the tower is actually listing itself with a split in the brickwork facing the river.

Gearing to Turbine Pump

photo: chris allen 2014

Six Mile House Mill, Halvergate
TG 461098   B25
River Bure, South Walsham Marshes

1797 Faden: 'Drain W. Mill'

1837 OS 1" sheet 47: X

1837 Cantley Tithe Map: two buildings shown; 1838 Apportionment: (233) 'Mill House, Yards &c'

1885 OS 6" sheet 66 SW: 'Draining Pump'

Listed grade II 5/50

Wailes: 10
Smith: 22

Downstream from the Stracey Arms Mill, sections of the River Bure's southern bank were detached parts of Burlingham and then Cantley parishes. Six Mile House sits on the boundary between the two and a little further on is this drainage mill.

photo: evelyn simak 2011

Wailes thus lists this mill as at Six Mile House on Cantley Marsh but with an incorrect map reference (that for Six Mile House Mill, Haddiscoe Island, Y27). He says it was rebuilt in the 1870's re-using much of the old gear.

Smith certainly includes it, describing a derelict tarred red brick tower with (then) four patent sails without the shutters and some internal gear. The list description confirms this and additionally mentions an external scoop wheel. As shown on the right in 1990 it then had but two sails remaining, and now without any must be a prime candidate for restoration.

photo: polystar 1990 N28

Kerrison's Level Mill, Halvergate
TG 462085   B26
River Bure, South Walsham Marshes

1837 OS 1" sheet 47:
'Marsh Mill' X

1838 Acle Tithe Map (third part): X
1840 Apportionment: —

1885 OS 6" sheet 66 SW:
'Draining Pump'

Listed grade II 5/51

Wailes: 77
Smith: 23

photo: evelyn simak 2011

This drainage mill is not immediately adjacent to any waterway as such. It sits remotely in the middle of the marshes about a mile north of The Fleet and a similar distance south of the River Bure, where parts of the southern bank are in Cantley and Runham parishes.

It is accordingly listed by Wailes as Key's Mill at Upper Runham, but lying south of both the Acle New Road and the parallel railway lines heading for Great Yarmouth, it might seem more akin to the mills on The Fleet. However the water it pumped would have found its way northwards along various dykes before discharging into the River Bure and its home parish of Acle is well and truly on the Bure, so it is included here.

The mill is of four storeys, the uppermost cylindrical one the result of heightening, the entire tower strapped with six wrought iron bands. The white weather-boarded boat-shaped cap has a surviving tail-pole adapted for winding by fantail. Internally the cast iron windshaft and horizontal drive to the scoop wheel survive, but not the connecting upright shaft.

Child's Mill, Runham
TG 471101   B27
River Bure, Runham Marsh (north)

1837 OS 1" sheet 47: X

1839 Runham Tithe Map: X
1840 Apportionment: (243) 'Mill &c.' owned and occupied by 'Commissioners of Drainage'

1885 OS 6" sheet 66 SW: 'Draining Pump'

Listed grade II 2/41 (as Runham Swim Windpump)

Wailes: 60 (63)
Smith: 54

photo: pierre terre 2008

The marshland immediately south of Runham village was drained by a squat mill on the north bank of the River Bure, reputedly built in the 1850's to replace an earlier mill slightly further east. It originally had an internal scoop wheel.

Smith describes a derelict tarred red brick tower, which at that time had only the lower boarding remaining on its cap, but a good deal of surviving machinery, hence its grade II listing. In the 1980's it was restored with a new white boarded boat-shaped cap and a fantail, but not sails.

Wailes seems to get confused here listing a 'Commission Mill' (60) built by a midland firm in 1819, which seems a good fit for this one, but also at (63) 'Runham Swim Mill' by Smithdale, both with a scoop wheel, and the same incorrect map reference.

Runham and Mautby Swims

A swim is the local term for a deep fording place where cattle could be coaxed across the river.

Villages such as Runham and adjoining Mautby had 'swims' which connected the two parts of the parishes, the northern settlement and the southern grazing meadows remotely positioned beyond the river.

Perry's Mill, Runham
TG 472099   B28
River Bure, Runham Marsh (south)

1837 OS 1" sheet 47: X

1839 Runham Tithe Map:
1840 Apportionment: —

1885 OS 6" sheet 66 SW: 'Draining Pump'

Listed grade II 4/42 (as Runham Windpump)

Wailes: 62
Smith: 53

Here on the southern bank of the River Bure and still in Runham parish, the adjoining area of grazing meadow also needed a drainage mill which was placed more or less opposite Child's Mill on the north bank. Wailes (62) describes this one as Perry's, Lake's or Upper Four Mile Mill and with a scoop wheel.

photo: pierre terre 2008

The modern OS map has the names Runham and Runham Swim interchanged compared to the listing descriptions, whilst Smith lists both as the latter, but sensibly qualified with (north) and (south).
This mill tower is taller than that opposite and has been made weather-tight with an aluminium cap on the original frame, protecting the internal gear that drove an external turbine, probably added later.

Lower Four Mile Mill, Mautby
(TG 489098)   B30
River Bure, Mautby Marsh (south)

A mile further down the River Bure from Perry's Mill there was a mill shown as X on the early OS map and as 'Draining Pump' in 1885. Faden in 1797 had shown two mills here as 'Drain W. Mills'.
Described by Wailes (61) as Lower Four Mile Mill, it was heightened and rebuilt before 1914, probably by Smithdale who most likely also upgraded the original common sails.

Five Mile House Mill, Runham
TG 478098   B29
River Bure, Runham Marsh (south)

1797 Faden: 'Drain W. Mill'

1837 OS 1" sheet 47: —

1839 Runham Tithe Map: X
1840 Apportionment: (317) 'Mill and Mill Drain'

1885 OS 25" sheet 66 SW: 'Draining Pump'

Listed grade II 4/39

Wailes: 59
Smith: 52

photo: polystar 1990 N30

Just downstream and not unlike Perry's Mill at Runham Swim, this one seems to have had the same temporary aluminium cap treatment to keep the weather out and protect its inner workings.

Wailes (59) describes the mill at this location as Brandford's Mill, with a turbine pump and attributes it to W.T. England. However to further complicate matters the presumed correct listing describes an external scoop wheel.

Smith, prior to the capping, observed a derelict tarred red brick tower without cap or sails, whilst the list description mentions a datestone inscribed 'N.M.N. 1849', presumably the date of some improvement works.

Another South Bank Pump
(TG 499091)
River Bure, Acle East Marshes

About a mile downstream from the site of Lower Four Mile Mill, there was another situated rather like the Stracey Arms Mill quite close to the Acle New Road and its parallel railway line heading into Great Yarmouth.
Not on the early maps, it appeared with a rectangular plan marked 'Draining Pump' on the 1885 6" OS map and was most likely engine driven.

Mautby Marsh Mill, Mautby
TG 489099   B31
River Bure, Mautby Marsh (north)

1797 Faden: 'Drain W. Mill'

1837 OS 1" sheet 47: X

1839 Mautby Tithe Map: X
1840 Apportionment: (189) 'Mill & Yard' owned by 'Robert Fellowes Esq'

1885 OS 6" sheet 66 SW: 'Draining Pump'

Listed grade II 4/40

Wailes: 46
Smith: 42

Across Mautby Swim from Lower Four Mile Mill, there was another mill that drained the northern part of Mautby Marsh, nearer the village, forming a further pairing of drainage mills across the waters of the Bure.

photo: polystar 1990 N31

Wailes (46) ascribes it a scoop wheel but gives no further detail, whilst Smith describes the usual for this part of the river, a derelict red brick tower without a cap in poor condition.

That situation was resolved in 1982 with its conversion to a holiday home. The stocks and fan gear are original, the cap and sails restorations and its sixteen sided upright shaft has been lost to make way for the new use.

Pump, opposite New Town
(TG 522097)
River Bure, Acle East Marshes

About a mile east Ashtree Farm Mill, the last mill still standing on the River Bure, another site drained the same marsh on its southern bank. This was at the eastern end of the last long bend in the river very close to Great Yarmouth.
Not on the early maps, it appeared on the 1885 6" OS map as 'Draining Pump', rectangular in plan and thus most likely engine driven.

Ashtree Farm Mill, Acle (detached)
TG 507095   B32
River Bure, Acle East Marshes

1837 OS 1" sheet 47: X

1838 Acle Tithe Map (fourth part): circular building shown
1840 Apportionment:
(758) 'Mill and Yard'

1885 OS 6" sheet 66 SW: 'Draining Pump'

Listed grade II 839-1/3/1

Wailes: 3
Smith: 4

Before the River Bure reaches Great Yarmouth there is one last big semi-circular sweeping bend about a mile in diameter. At its western end the mill at Ashtree Farm drained the marshland enclosed south of the river, belonging to the parish of Acle.

photo: pierre terre 2008

Wailes attributes this mill to millwright Thomas Smithdale of Acle and mentions a 1912 rebuild.

At the time of Smith's survey it had a bit more of the cap still in place than in 1994 when the windshaft, brakewheel and scoop wheel were all exposed to the elements.

Now fully restored, most of the internal gear remains including a timber vertical shaft and quite a lot of cast iron components leading to the newly cased scoop wheel. A good example of what sensitive restoration can achieve.

photo: polystar 1994 S11

Six West Caister Sites

River Bure, West Caister Marshes

Three Mile House Mill, West Caister
(TG 506096)   B33
River Bure, West Caister Marshes

Six sites at one time or another drained the large area of West Caister Marshes north of the River Bure where it performed its last big turn. The first was part of another pairing of mills, situated on the north bank opposite the mill at Ashtree Farm.

This tower mill appeared as X on the 1837 1" OS map and as a small circular building plan within a semi-circular drain on the undated Caister Tithe Map. The 1843 Apportionment lists (397) nearby as 'Upper Mill Marsh' and the mill later appears as 'Draining Pump' on the 1885 6" OS.

Hollow Post Mill, West Caister
(TG 507111)   B34
River Bure, West Caister Marshes

About a mile due north of Three Mile House Mill, another 'Draining Pump' is recorded by the 1885 6" OS map, right on the northern edge of the marshland, immediately south of the village of West Caister.

This was a hollow post mill with a plunger pump, listed by Wailes (99) as constructed by Whitmore & Binyon, and mentioned by Smith in his 'Some Old Sites' list. Not on the earlier maps, this was obviously a later 19th Century addition to the mills in this area.

Trestle Mill, West Caister
(TG 512102)   B35
River Bure, West Caister Marshes

Further round the big bend on the River Bure, Faden in 1797 shows a 'Drain W. Mill', which Wailes (8) includes as a tower mill with 'No notes'. Smith lists a steel trestle mill here that drove a turbine pump, which Wailes (88) attributes to Smithdale.
This one did appear on the early maps, as X on the 1837 1" OS map and as a circular enclosure on the undated West Caister Tithe Map. The 1843 Apportionment lists nearby 'Mill House Marsh', but interestingly at (368) 'Steam Engine Yard'. This is an early introduction of such an engine, presumably made possible by the proximity of coal via Great Yarmouth.

Further Sites, West Caister
(TG 509101)
River Bure, West Caister Marshes

Between Three Mile House Mill and Trestle Mill on the great arc of the River Bure, there was in the later 19th Century another West Caister site. Not on the early maps this one, like the others, appeared as 'Draining Pump' on the 1885 6" OS map, along with two others shown at TG 523110 & 525105.

From here onwards downstream the River Bure passes behind Great Yarmouth before discharging into Breydon Water, where its waters mingle with the those of the Rivers Yare and Waveney, the first of which we will next consider.

2.4 Upper River Yare Drainage Windmills

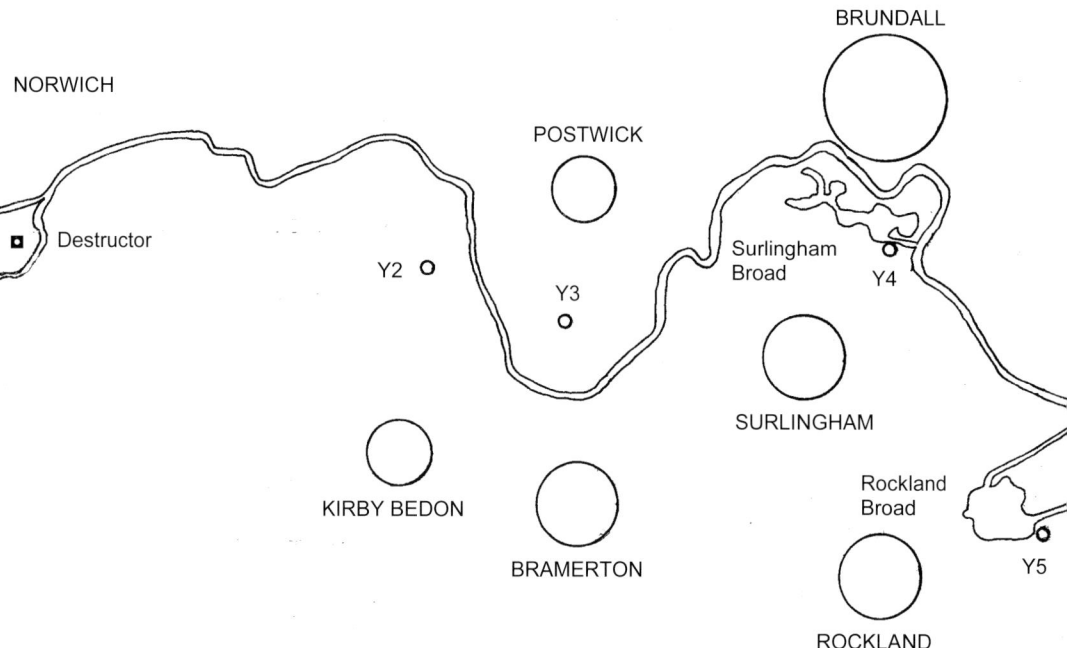

River Yare Drainage Mills

We now have to go back west again and start the River Yare in its course eastwards from Norwich to Great Yarmouth, again taking it in three stages.

Unfortunately our first site at Eaton (Y1) lies too far west of Norwich to show here, but the timber yard destructor described opposite just squeezes in.

Scale: 1 inch to 1 mile

Timber Yard Destructor, Thorpe St Andrew
TG 248075
River Yare

1838 OS 1" sheet 46: —

1938 OS 25" sheet 63:
two circular plans marked 'Destructors'

Listed grade II 4/10038

Wailes: —
Smith: —

photo: polystar 2016

This tapering red brick tower can be glimpsed enigmatically from the train as one approaches Norwich Station, but not always, dependent on the positions of trucks and piles of materials there in the aggregate works next to the line.

The view is a recent one as formerly the large sheds of a timber yard on the south bank of the River Wensum intervened. The circular building sits on a spit of land between that river and the River Yare, alongside which the maps reveal a network of drainage ditches south of the structure, but without any indication of a draining pump.

A single circular plan first appears on the 1928 25" OS map without any identification, joined by a second and the word 'Destructors' on the 1938 map. Two circles remained in 1956, but only the eastern one is there now.

It is fortunately listed grade II, the description giving it as a 'bottle kiln' because of its shape, but the listing leaves it unclear as to what was actually burnt. The proximity of the timber yard suggests it was actually timber waste that was consumed and destructors were sometimes used to generate electricity. This in its turn may well have driven more recent drainage pumps locally, so not a drainage mill after all, but included for its interest.

Four Upper Yare Sites

River Yare, Various Marshes

Eaton Mill, Eaton
(TG 198068)　　　Y1
River Yare, Eaton Marsh

The River Yare that comes from further west and sweeps around the south side of Norwich appears to have no drainage mills around the city's southern edge. The first encountered coming downstream is however just south-west of the city at what was once the separate village of Eaton.

Although nothing is shown on the early 1838 1" OS sheet 46 nor on the Eaton Tithe map , the 1885 6" OS map sheet 63 SW clearly shows 'Draining Pump (Wind)' on the river's edge here.

Kirby Bedon Mill, Kirby Bedon
(TG 283072)　　　Y2
River Yare, Kirby Marsh

Very nearly as far west as the sites found at the head of the River Bure, the first site found coming down the River Yare below Norwich is at Kirby Bedon, where the remains of a pump were found in Whitlingham Country Park. The mill there drained Kirby Marsh into a dyke that took the water about a third of a mile north-eastwards to the river proper.

It appeared on the 1838 1" OS sheet 46 with the usual X symbol, so was definitely wind powered, appearing later in 1887 on the 6" OS map sheet 64 SW as 'Draining Pump'.

Postwick Marsh Mill, Postwick
(TG 297068)　　　Y3
River Yare, Postwick Marsh

Beyond Kirby Bedon Mill the River Yare takes a nearly 180° turn to the left forming a large sweeping arc enclosing Postwick Marsh and village on its northern side.

In the centre of this marsh another X symbol is to be found on the 1838 1" OS map sheet 46, indicating a wind driven mill. A similar symbol also appears here on the 1838 Postwick Tithe Map, which is described in the 1840 Apportionment (207) as 'Bank & Mill'. By 1887, however, this mill had gone.

Another Site, Postwick
(TG 302070)
River Yare, Postwick Marsh

Not on the early OS or Tithe maps the eastern part of Postwick Marsh was drained at a point on the river bank south-east of the village itself, where further pump remains were found. Here another 'Draining Pump' is indicated by the 1887 6" OS map sheet 76 NW, shown along with a rectangular building, most likely containing an engine driven pump.

None of these upper Yare sites appear to have been included in the listings given by Wailes or Smith, until we get down the river as far as Claxton (Y6).

More Upper Yare Sites

River Yare, Various Marshes

West End Pump, Surlingham
(TG 300063)
River Yare

On the south-east bank of the River Yare opposite Postwick Marsh the western part of Surlingham parish forms the river bank as the river readies itself for another 180° turn back to the right, enclosing this time the village of Surlingham on the southern side.

Here there was a 'Draining Pump' shown on the 1887 6" OS map sheet 76 NW, but nothing was shown on the earlier OS or tithe maps. This is therefore likely to be either engine driven or possibly a trestle type mill, since no building outline is shown.

Pump, Postwick
(TG 303074)
River Yare, Postwick Marsh

Set back a short distance from the north bank of the River Yare as it approaches the bend around Surlingham another 'Draining Pump' was shown on the 1887 6" OS map sheet 64 SW with a rectangular building footprint. It also did not appear on the earlier maps, probably indicating another later engine driven pump.

About half a mile beyond this site another is shown in Postwick (TG 309079), where the 1887 25" OS map has 'Pump' alongside industrial buildings shown as 'Grease Manufactory', so probably not connected to marshland drainage.

Church Pump, Surlingham
(TG 304071)
River Yare

On the southern bank of the River Yare, below the site of Surlingham's old Church and roughly opposite the second and third Postwick sites above right, another 'Draining Pump' is indicated by the 1887 1" OS map sheet 64 SW.

Again with no earlier reference such as the 1838 1" OS map or the 1839 Tithe Map, which are both blank, this has to be assumed probably not wind powered or at best a trestle type mill as no building plan is shown in 1887.

Coldham Mill, Surlingham
(TG 333074)   Y4
River Yare, Surlingham Marsh

Further round the bend and north-east of Surlingham village a 'Draining Pump' is indicated by the 1887 OS maps near Coldham Hall.

The larger scale 25" map appears to show a circular building plan here so this is most likely a tower mill, although there is nothing to confirm it as such from the earlier OS or Tithe Maps.

It would have drained an area known as The Outmeadows east of Surlingham Broad.

Even More Upper Yare Sites

River Yare, Various Marshes

Pump, Surlingham
(TG 327061)
River Yare, Surlingham Marsh

Opposite Strumpshaw Marsh on the south-west bank of the River Yare, there is a mile long stretch of Surlingham Marsh running south from Coldham Hall.

Here, near Grange Farm at the back edge of the marsh there was a 'Draining Pump (Disused)' shown on the 1887 6" OS map sheet 76 NE.

Shown with a rectangular building plan, this was most likely another engine driven pump.

Strumpshaw Pump, Strumpshaw
(TG 331066)
River Yare, Bradeston Marsh

Bradeston is a small parish comprising little more than Braydeston Hall and its church, now absorbed into the greater parish of Brundall on the north bank of the River Yare. Bradeston Marsh adjoins the river and had strips of marsh interleaved with those of Strumpshaw, the next parish to the east.

Not on 1846 Strumpshaw Tithe Map, another 'Draining Pump (Disused)' is shown on the 1887 6" OS map sheet 76 NE along with a rectangular building plan.

Long's Corner Mill, Rockland
(TG 336049)   Y5
River Yare, Rockland Marsh

Beyond the southern end of Surlingham Marsh lies Rockland Broad, at the south-east corner of which was Long's Corner. Here the 1887 6" OS map sheet 76 NE shows another 'Draining Pump', with a circular building plan, indicative of a wind driven mill.

Although there is nothing to confirm this on the earlier 1838 1" OS map nor the 1840 Rockland Tithe Map, an old postcard from 1908 reliably shows a trestle mill there. The remains of a turbine pump casing have recently been found there.

Strumpshaw Pumping Station, Strumpshaw
TG 341057
River Yare, Strumpshaw Marsh

1838 OS 1" sheet 46: —

1846 Strumpshaw Tithe Map: —

1887 OS 6" sheet 76 NE: —

1907 OS 25" sheet 76: 'Draining Pump', with large rectangular building plan

Wailes: —
Smith: —

Not on the site of an earlier mill as attested to by its absence from the early maps, this pumping station does not appear on the OS 25" maps until 1907, so it can be dated as no earlier than the previous 1887 edition.

The apparatus here probably superseded the upstream and downstream pumps on the northern bank of the Yare and drained most of Strumpshaw Marsh.

The station comprises a large white brick engine shed with a now free-standing white brick octagonal chimney on a square base and was variously powered by steam, diesel and electricity. Like Black Mill on the Waveney, this notable local landmark should perhaps be listed.

photo: evelyn simak 2009

Common Pump, Strumpshaw
(TG 342052)
River Yare, Strumpshaw Common

Just over a mile further down river from Strumpshaw Pump another 'Draining Pump (Disused)' was shown on the 1887 6" sheet 76 NE.

Like the other Strumpshaw pumps, it too did not appear on the 1846 Strumpshaw Tithe Map or earlier OS maps and was probably also engine driven.

Middle River Yare Drainage Windmills

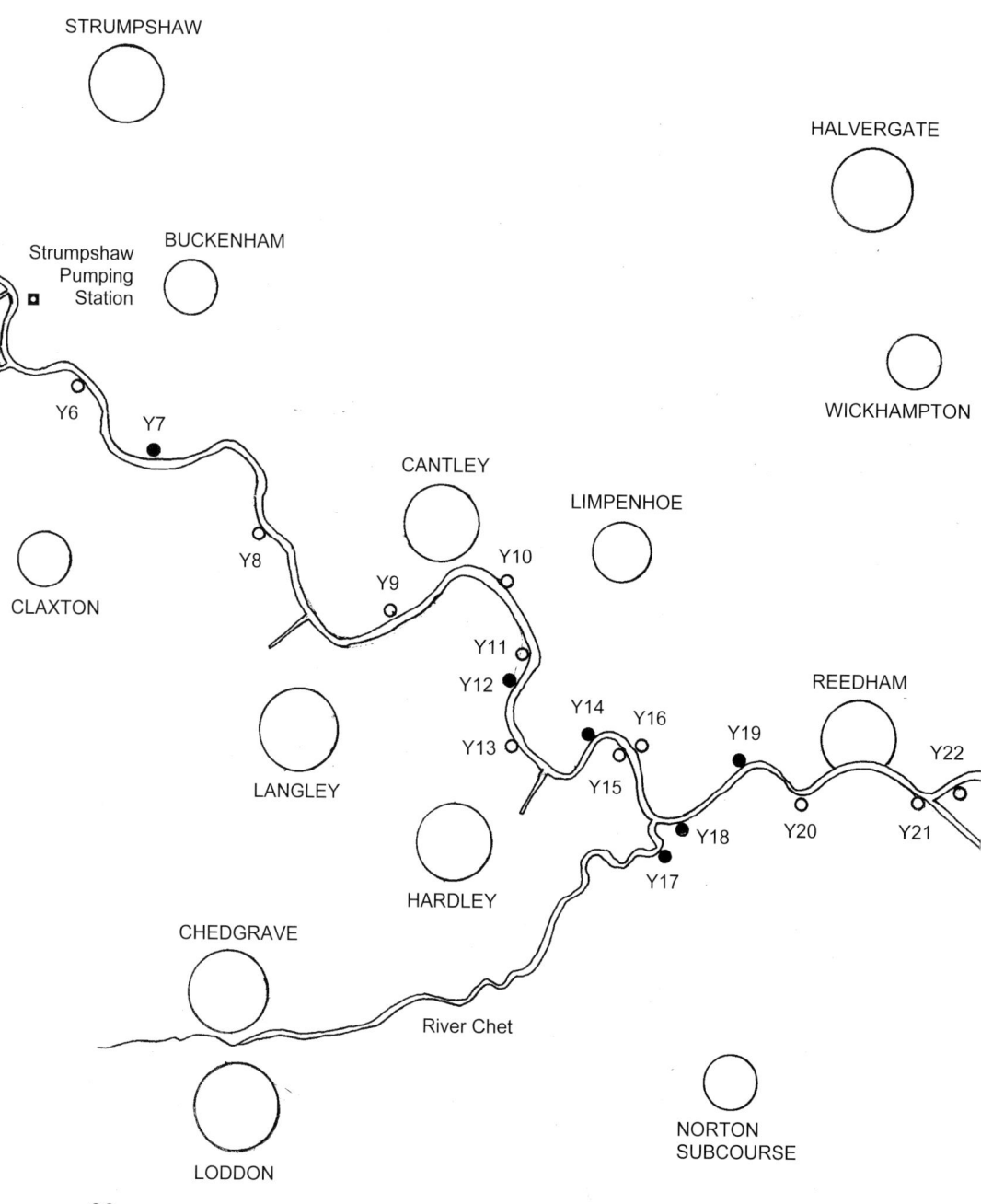

Buckenham Ferry Mill, Buckenham
TG 353044   Y7
River Yare, Buckenham Marsh

1797 Faden: circular plan

1838 OS 1" sheet 46: X

n.d. Buckenham Tithe Map: X; 1838 Apportionment: (101) 'Mill and Yard'

1887 OS 6" sheet 76 NE: 'Draining Pump (Wind)' and 'Draining Pump' adjacent

Wailes: 7
Smith: 11

photo: polystar 1994 S25

This three storey brick tower mill is relatively early, having appeared on the early 19th Century OS and Tithe maps. It drained a large area of marshland on the north bank of the River Yare immediately south of the small village of Buckenham.

Wailes just says of it 'No notes' and unfortunately assigns it an incorrect map reference. Smith describes the tower as derelict and slightly leaning with a conical roof where the cap would have been. Whilst the tower survives, it now has a flat roof and the lean-to Smith describes (which may have housed an auxiliary pump) has no roof at all.

Claxton Mill
(TG 346051)   Y6
River Yare, Claxton Marsh

About half a mile downstream from 'Short Dike' leading to Rockland Broad, a long dyke cuts across Claxton Marsh on the south bank of the River Yare.
Not on the early maps, 'Draining Pump' appears on the 1887 6" OS map, but as a rectangular plan building within a semi-circular ditch, suggesting something other than wind power at that time. Wailes however (13) speaks of sails by W.T. England and Smith lists a tower mill.

Marsh Farm Mill, Cantley
(TG 376030)   Y9
River Yare, Cantley Marsh

1838 OS 1" sheet 46: X

1837 Cantley Tithe Map: —
1838 Apportionment: (197) 'Mill Marsh'

1885 OS 6" sheet 77 SW: 'Draining Pump'

Wailes: 9
Smith: Some Old Sites

Cantley Marsh, south-west of the village, was drained by a long dyke and a mill on the north bank of the River Yare. Wailes says the mill drove a turbine pump and gives a date of 1874, whilst Smith describes 'a few bricks and timbers left', all now replaced by this fairly typical modern replacement brick pump-house.

photo: polystar 2016

Langley Green Mill, Langley
(TG 363037)   Y8
River Yare, Langley Marshes

Still shown as 'Draining Pump' on modern OS maps, the 1887 6" map shows a rectangular building at this site, suggesting an engine shed at that time.

In the absence of earlier map evidence, there is only nearby 'Mill Dike' to suggest a wind powered drainage mill at one time worked here.

Sugar Factory Mills, Cantley
(TG 386033 & 391036)   Y10
River Yare, Cantley Marsh

South-east of Cantley village the 1885 6" OS map shows a 'Draining Pump' at two further sites near the sugar factory. That to the north has a rectangular building.

Both the riverside site and Marsh Farm Mill were adjoined by a 'Mill Marsh' on the 1837 Cantley Tithe Map. Both appeared on the 1838 1" OS sheet 46 with a symbol X and on the 1885 6" OS as 'Draining Pump'.

Hardley Mill, Hardley
TG 388024   Y12
River Yare, Hardley Marshes

1838 OS 1" sheet 46: —

1840 Hardley Tithe Map: —
1841 Apportionment: —

1885 OS 6" sheet 77 SW: 'Hardley Draining Pump (Wind)'

Listed grade II 3/41

Wailes: 37
Smith: 35

photo: polystar 2016

Draining the very northern edge of Hardley Marshes, this mill is described by both Wailes and Smith as that for Langley Marshes, which adjoin to the north. Further confusion is added by the list description which cites it as Hardley Windpump, Langley.

Smith describes a derelict red brick tower with a collapsed boat-shaped cap and no sails, however sufficient internal gear survives to warrant its grade II listing. This includes the chamfered square vertical shaft with cast iron wallower (see p.12) and spur wheel, leading to a turbine chamber.

The tower is dated 'TWBPB 1874' and the whole mill has now been fully restored.

Langley Mill, Langley
(TG 389026)   Y11
River Yare, Langley Marshes

A short distance north of Hardley Mill and actually draining Langley Marshes, there was another mill site on the 1885 6" OS map, by then a rectangular building described as 'Langley Draining Pump'.
This was definitely earlier than Hardley Mill, as unlike that one it had appeared on Faden's 1797 map as 'Drain Mill' and the 1838 1" OS map as 'Langley Mill' with the usual X symbol.

Limpenhoe Mill, Limpenhoe
TG 395019   Y14
River Yare, Limpenhoe Marshes

1838 OS 1" sheet 46: X

n.d. Limpenhoe Tithe Map: circular plan shown
1846 Apportionment (140a) 'Mill & Yard'

1885 OS 6" sheet 77 SW: 'Draining Pump'

Listed grade II 6/22

Wailes: 35
Smith: 36

photo: polystar 2016

Although on the opposite side of the river from Langley parish, this mill is listed as 'Langley No.3, Cantley', a name that seems to derive from Wailes to distinguish it from his Langley Marsh (37) and Langley Detached (36) listings.

Another derelict red brick tower, it has four storeys and does retain some remains of the cap along with a good deal of internal machinery. This includes the upright shaft, a 10 foot diameter pit wheel and horizontal drive shaft to an external 20 foot diameter scoop wheel.

The cast iron components are by W.T. England of Great Yarmouth and dated 1895, although the mill itself is earlier 19th Century.

Hardley Street Mill, Hardley
(TG 388018)   Y13
River Yare, Hardley Marshes

The 1838 1" OS sheet 46 has a symbol X at this site annotated as 'Hardley Mill'.

Thought at first to be a potential mis-positioned symbol, it is ratified by the 1840 Hardley Tithe Map which has a small enclosure here described in the 1841 Apportionment (76) as 'Mill & Yard'. An early mill site then, now lost and not appearing on later maps.

Boyce's Dyke Mill, Norton
TG 401008   Y17
River Chet, Norton Marshes

1838 OS 1" sheet 46: —

n.d. Norton Subcourse Tithe Map: —
1841 Apportionment: —

1887 OS 25":
'Draining Pump'

Wailes: —
Smith: 44

photo: polystar 2016

Draining the very western edge of Norton Marshes, this mill actually pumped its water from 'Mill Dyke' into Boyce's Dyke near where it joined the River Chet, a short distance above the latter's junction with the south side of the River Yare.

Wailes seems to have missed this one, but Smith describes a derelict truncated tarred red brick tower with but two storeys remaining, topped by a conical and apparently slated roof (according to his photograph), which has now gone.

Empty apart from a single floor accessed by steps, it has an adjoining corrugated iron clad structure, which might be related to some later pumping use.

Another Mill, Hardley
(TG 398018)   Y15
River Yare, Hardley Marshes

The eastern part of Hardley Marshes, opposite Limpenhoe Mill, was drained by yet another mill in the parish of Hardley.
The site appeared on both the 1838 1" OS map with the usual X symbol and on the 1840 Hardley Tithe Map, being listed in the 1841 Apportionment (115) as 'Mill & Yard'. However it did not later appear on the 1885 6" OS map.

Norton Marsh Mill, Norton
TG 403011   Y18
River Yare, Norton Marshes

1838 OS 1" sheet 46: —

n.d. Norton Subcourse Tithe Map: circular plan only
1841 Apportionment: —

1885 OS 6" sheet 77 SW: 'Draining Pump'

Wailes: 25
Smith: 43

photo: polystar 2016

Although clearly east of the River Chet's junction with the River Yare and thus in Norton parish, Wailes lists this one as in 'Hardly' (sic). He attributes it to millwright D. England and says it drove a turbine pump.

It did however have a very visible scoop wheel housing adjoining the three storey tarred red brick tower, as shown on page 116 in 1994 along with the remains of its earlier novel conical metal sheeted roof.

Now fitted with a new boat-shaped cap, it has been converted into living accommodation and is not listed, probably on account of the loss of internal gear that entails. A small modern red brick pump house adjoins.

Another Mill, Limpenhoe
(TG 399019)   Y16
River Yare, Limpenhoe Marshes

Directly over the river from the last mill described in Hardley (Y15) another helping to drain Limpenhoe Marshes was shown by Faden in 1797 as 'Drain Mill'.

It appeared on both the 1838 1" OS map with the usual X symbol and on the undated Limpenhoe Tithe Map, being listed in the 1846 Apportionment (144) as 'Mill Yard'. However its appearance on the 1885 6" OS map as 'Draining Pump' adjoins a rectangular building.

Reedham Ferry Mill, Reedham
TG 409017　　　Y19
River Yare, Reedham Marshes (west)

1838 OS 1" sheet 46: —

1841 Reedham Tithe Map: —

1841 Apportionment: —

1885 OS 6" sheet 77 SW: 'Draining Pump'

Wailes: 53
Smith: 47

photo: polystar 1994 S23

Also known as 'Red Mill' with the brickwork now painted to suit, this former drainage mill has been converted for residential use with the addition of a wrap around cantilevered first floor conservatory, which does unsettle the mill aesthetic a little.

Wailes attributes this drainage mill with a turbine pump to millwright D. England, and with its cylindrical top storey it does appear to have been heightened, probably late in the 19th Century. Smith finds it already converted in 1974, so without any early map evidence, its time as a working drainage mill was perhaps less than a century.

Another Mill, Norton
(TG 415013)　　　Y20
River Yare, Norton Marshes

Wailes (54) mentions another mill at Reedham Ferry, but on the south bank at TG 408014, for which no map evidence can be found. He lists it as by Barnes and with a scoop wheel.
Half a mile east of there, another mill site is shown by Faden in 1797 'Norton Drain Mills' and also appears on the 1838 1" OS map with a mill symbol X, confirmed by the Norton Subcourse Tithe Map for which the 1841 Apportionment (35) lists 'Mill and Yard'.

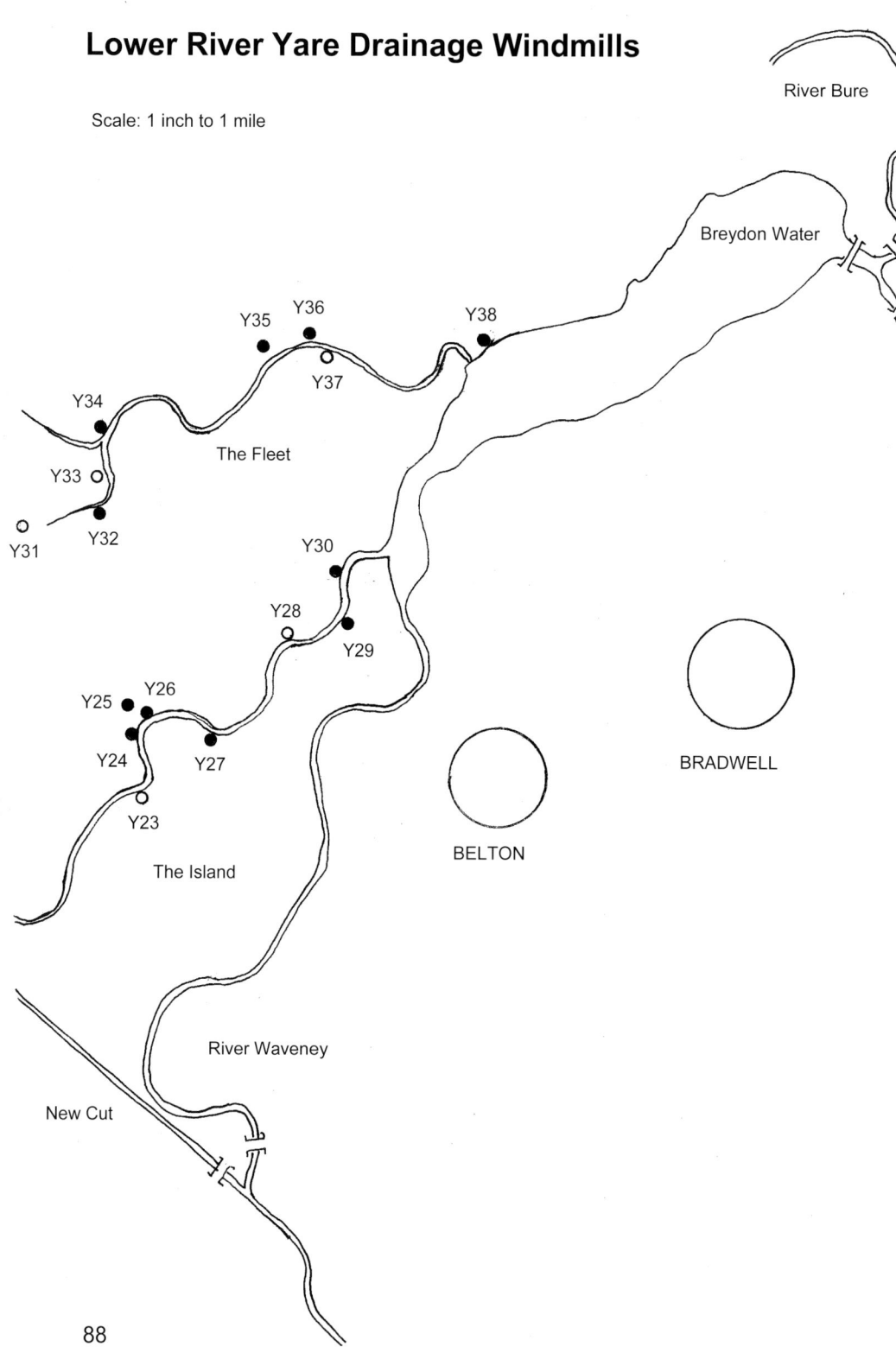

Polkey's Mill, Reedham
TG 444035   Y24
River Yare, Reedham Marshes

1797 Faden: 'Drain W. Mills'

1837 OS 1" sheet 47:
X 'Reedham Mills'

1841 Reedham Tithe Map: circular plan only
1841 Apportionment: (80) 'Cottage Mill and Garden'

1885 OS 6" sheet 77 SE: 'Draining Pumps'

Listed grade II* 7/71

Wailes: 56
Smith: 49

photo: evelyn simak 2011

One of three sites known as Reedham Mills, this one is the southernmost, located near Seven Mile House on a sharp bend in the River Yare. The 1885 6" OS map uses the plural as an engine house adjoins it.
Wailes thus lists it as 'South Mill', giving a map reference a little too far south. Smith corrects this, describing a derelict tarred red brick tower without cap, a situation later remedied by a full restoration. It has a scoop wheel and enough of its stocks and internal gear remain to justify its listing at the higher grade of II* with the additional credit of 'Group Value'.

New Cut Mill, Raveningham (det)
(TG 426013)   Y21
River Yare, Norton Marshes

Just where the New Cut diverges south-eastwards off the Yare, a further mill in a detached portion of Raveningham helped drain the eastern end of Norton Marshes. Faden showed two circular symbols here in 1797, part of three 'Norton Drain Mills'. The Raveningham Tithe Map has (20) 'Mill Yard' and the early OS maps show here both a mill symbol X in 1838 and 'Draining Pump' in 1885. Now it is simply marked 'Pump House'.

North Mill, Reedham
TG 443036   Y25
River Yare, Reedham Marshes

1837 OS 1" sheet 47:
X 'Reedham Mills'

1841 Reedham Tithe Map:
semi-circular ditch only
1841 Apportionment: (82) 'Mill and Yard'

1885 OS 6" sheet 77 SE: 'Draining Pump'

Wailes: 55
Smith: 50

photo: ashley dace 2010

Dubbed 'North Mill' by Wailes and set away from the River Yare, this derelict shell of a drainage mill is one of the three Reedham Mills. Two appear as 'Draining Pump' on the 1885 6" OS map with circular building plans, whilst Polkey's Mill has a rectangular plan too.

Wailes says this one drove a scoop wheel and his map reference for this one is also a little too far south. Smith again corrects the map reference and talks correctly of a ruined empty shell, leaning over.

Not listed on account of its poor condition and lack of gear, it is nevertheless mentioned in the other listings hereabouts as it provides part of their Group Value.

Another Mill, Reedham
(TG 429014)   Y22
River Yare, Reedham Marshes (south)

Although not shown on the early OS map, Faden in 1797 showed 'Drain W. Mill' here. The 1841 Reedham Tithe Map shows a circular ditch at this site on The Island, just below the New Cut junction, and The 1841 Apportionment (456) lists it as 'Mill and Yard' owned and occupied by 'John Stracey'. It later appears on the 1885 6" OS map as a rectangular building within a circular ditch.

Cadge's Mill, Reedham
TG 446036   Y26
River Yare, Reedham Marshes

1797 Faden: 'Drain W. Mills'

1837 OS 1" sheet 47:
X 'Reedham Mills'

1841 Reedham Tithe Map:
circular plan only
1841 Apportionment:
(79) 'Mill Marsh' adjoining

1885 OS 6" sheet 77 SE:
'Draining Pump'

Listed grade II 7/70

Wailes: 51
Smith: 48

photo: colin mitchell 2006

Of the group of three sites known as Reedham Mills, Cadge's is the easternmost, located near Seven Mile House, but further round the sharp bend in the River Yare. This group were replaced first by steam and then by diesel in 1941 as shown below.

Wailes lists this one as 'Lower Seven Mile Mill' or 'Cadge's Mill', attributing it to millwright Smithdale.

Smith renders the usual description of a derelict tarred red brick tower without a cap or sails, but retaining a windshaft, brakewheel and an internal 18 foot diameter scoop wheel. The upper parts of this had been removed by 1994, and were later reinserted within a new cap.

photos: evelyn simak 2011

Six Mile House Mill, Haddiscoe Island
TG 452033　　　Y27
River Yare, Chedgrave Marshes

1797 Faden: 'Drain W. Mill'

1837 OS 1" sheet 47: X

1839 Chedgrave Tithe Map: X;　1839 Apportionment: (178) 'Mill Yard'

1885 OS 6" sheet 77 SE: 'Draining Pump'

Listed grade II 4/11

Wailes: 11
Smith: 15

Described by Wailes as 'Steven Hewitt's Mill', this mill drained with its scoop wheel the northern part of a detached portion of Chedgrave parish on Haddiscoe Island, half a mile downstream from the three Reedham Mills.

photo: polystar 1994 S20

The tithe information includes Francis Perkins as owner and Edward Hewitt as occupier, which collaborates the name given. The modern OS map has it still as 'Draining Pump (disused)'.

Although it has lost its cap, the windshaft and brakewheel survive atop the three storey tarred red brick tower, which contains a flue within its wall thickness. Inside there is an octagonal vertical shaft with a cast iron wallower.

Upper Seven Mile Mill, The Island
(TG 446028)　　　Y23
River Yare, Chedgrave Marshes

Half a mile upstream and on the opposite bank of the River Yare from Reedham Mills, Faden shows in 1797 a 'Mill' draining this part of Haddiscoe Island belonging to Chedgrave parish. Wailes (52) attributes it to millwright Barnes with yet another scoop wheel and an incorrect map reference, whilst Smith lists it in his 'Some Old Sites', correcting its former position as above.

Langley Detached Mill, Haddiscoe Island
TG 466045   Y29
River Yare, Langley Marshes

1797 Faden: 'Drain W. Mill'

1837 OS 1" sheet 47: X

1841 Langley Tithe Map: —
1841 Apportionment: —

1884 OS 6" sheet 78 SW: 'Windmill (Pumping)'

Listed grade II 4/17

Wailes: 36
Smith: 34

photo: pierre terre 2008

This drainage mill sits near Raven Hall in the detached portion of Langley Marshes. It is almost at the sharp end of The Island, about half a mile before the River Yare is joined by the River Waveney to form Breydon Water.

In 1994 'Red Mill' might have been a better name, since the white boat-shaped cap on this four storey red brick tower was then unusually painted red with the appearance of rusty corrugated iron (see p.116).

The mill drove a scoop wheel and retains its curb and track inside the cap, along with the windshaft and brakewheel. However, internally the rest is probably lost as it has been converted with the addition of a small thatched extension.

Tuck's Mill, Reedham
(TG 459044)   Y28
River Yare, Reedham Marshes

A mile and a half downstream from Reedham Mills this mill helped drain the eastern part of Reedham Marshes. It appeared on the 1841 Reedham Tithe Map as a circular building plan, listed (53) as 'Mill Yard and Rand'. OS maps show it as X on the 1837 1" and 'Draining Pump' on the 1885 6".
Wailes (5) describes it with a scoop wheel, whilst Smith lists it in his 'Some Old Sites'.

High Mill, Berney Arms
TG 465049   Y30
River Yare, Berney Marshes

1797 Faden: 'Drain W. Mill'

1837 OS 1" sheet 47: 'Sawing Mill'

1841 Reedham Tithe Map:
1841 Apportionment: (32) 'Factory Mill and Yard'

1884 OS 6" 78 NW: 'Windmill (Pumping)'

Scheduled Monument

Wailes: 4
Smith: 9

photo: polystar 1994 S18

This drainage mill and the adjoining Berney Arms Inn are accessible only by boat from the River Yare or by footpath from the nearby lonely railway station. Scheduled and in the care of English Heritage, it is perhaps better protected than any mere listed mill. Without a list description it is difficult to compare with the others, but at seven storeys it is certainly the tallest drainage mill in the country.

Wailes attributes it to millwright Stolworthy in 1865, which was a rebuild replacing an earlier mill, adding the grinding of cement clinker to its duties. Smith reports it in good working order, with a tarred red brick tower topped by the usual white boarded boat-shaped cap and sporting a large external scoop wheel.

Wickhampton Mill, Wickhampton
(TG 433054)   Y31
The Fleet, Wickhampton Marshes

The marshes north-west of Berney Arms are drained by a tributary of the River Yare, The Fleet. At the western edge of this area Faden in 1797 shows 'Drain W. Mill' near Wickhampton village.
This was shown with a mill symbol X on both the 1838 1" OS map and the undated Wickhampton Tithe Map from around the same time.

Stone's Mill, Wickhampton
TG 441056   Y32
The Fleet, Wickhampton Marshes

1797 Faden: 'Drain Wind Mill'

1837 OS 1" sheet 47: —

n.d. Wickhampton Tithe Map: —
1843 Apportionment: —

1885 OS 6" sheet 77 NE: 'Draining Pump'

Listed grade II 4/28

Wailes: 23
Smith: 24

photo: evelyn simak 2007

About a mile east of Wickhampton village adjoining The Fleet, this drainage mill, attributed by Wailes to millwright Barnes, comprises the usual derelict tarred red brick tower of four storeys without a cap.

According to Smith the tower is leaning and it was also known as 'Kerry's Mill'.

However a good deal of the cap's contents remain including rare spur gear winding, the cast iron windshaft and brakewheel, along with curb and tracks etc. Down below there is a cast iron scoop wheel shaft.

Carter's Mill, Halvergate
(TG 441059)   Y33
The Fleet, Halvergate Marshes

A little north of Stone's Mill, there was an earlier one listed by Wailes (19) and Smith in his 'Some Old Sites' as Carter's Mill (or Kerry's Mill), near what is now called Marshman's Cottage.
Shown as X 'Mill' on the 1837 1" OS map, it appears on the 1839 Halvergate Tithe Map, given in the 1842 Apportionment (262) as 'Mill and Garden' on Cantley Level, and then later on the 1885 6" map as 'Draining Pump'.

Mutton's Mill, Halvergate
TG 441064   Y34
The Fleet, Halvergate Marshes

1797 Faden: 'Drain W. Mill'

1837 OS 1" sheet 47: 'Mill' X

1839 Halvergate Tithe Map: circular building plan
1842 Apportionment: (267) 'Mill and Garden'

1885 OS 6" sheet 77 NE: 'Draining Pump'

Listed grade II* 4/44

Wailes: 21
Smith: 25

About half a mile north of Stone's Mill and also alongside The Fleet that here veers eastwards, there was a mill draining Frothelme's Level on Halvergate Marshes. Wailes attributes the original four storey mill here to Stolworthy of Great Yarmouth.

photo: polystar 1994 S16

It had been described by Smith as yet another 'derelict tarred red brick tower without cap', but this has fortunately now been remedied. Restored in 1980, this grade II* listed drainage mill is described as the only surviving example of an internal scoop wheel in the Broads area.

Now fitted with two sails, its contrasting white boarded cap and tarred tower show well what many other disused mills could look like. Internally it does retain a cast iron windshaft, wallower and a timber drive shaft along with a cast iron crown wheel and pit wheel that drove the scoop wheel.

Another Mill, Halvergate
(TG 426069)
Halvergate Marshes

The 1885 6" OS map shows a 'Corn Mill & Draining Pump' at this site just east of Halvergate village.

Shown as a rectangular building, it does not appear on the earlier OS or Tithe maps.

High's Mill, Halvergate
TG 457071   Y35
The Fleet, Halvergate Marshes

1797 Faden: 'Drain W. Mills'

1837 OS 1" sheet 47:
'Mill' X

1839 Halvergate Tithe Map:
circular building plan
1842 Apportionment: (361)
'Mill and Garden'

1885 OS 6" sheet 77 NE:
'Draining Pump'

Listed grade II 5/43

Wailes: 20
Smith: 26

A mile and a half further down The Fleet from Mutton's Mill, High's Mill drained Colman's Level on the eastern part of Halvergate Marshes. Wailes notes that this drainage mill was also known as 'Gilbert's Mill'.

photo: polystar 1994 S15

Meanwhile Smith describes the usual 'derelict tarred red brick tower in poor condition without cap' adding that it had 'windshaft, brakewheel gear inside, scoop wheel outside'.

It does now benefit from a temporary aluminium cap over the cap frame, protecting the surviving internal gear, so that future restoration along the lines of Mutton's Mill is at least feasible.

The Fleet

Essentially a short tributary of the River Yare, The Fleet drains a large expanse of Halvergate Marshes between the Rivers Bure and Yare. It flows eastwards from near Wickhampton, discharging into Breydon Water just east of Lockgate Mill.

Two groups of drainage mills feed into it, that in the west includes Stone's and Mutton's Mills, the other in the middle of the marshes includes High's and Howard's Mills.

Howard's Mill, Halvergate
TG 462072   Y36
The Fleet, South Walsham Marshes

1797 Faden: 'Drain W. Mills'

1837 OS 1" sheet 47: 'Mill' X

1841 South Walsham Tithe Map: not available
1842 Apportionment: —

1884 OS 6" sheet 78 NW: 'Windmill (Pumping)'

Listed grade II 5/52

Wailes: 22
Smith: 27

photo: polystar 1994 S14

Sited on a detached portion of South Walsham parish, this mill is about a quarter mile further down The Fleet from High's Mill. Also known as South Walsham Mill it drained the southern part of South Walsham Marshes.

Wailes attributes this mill to millwright Barnes, mentioning a rebuild by W.T. England. Smith has it as derelict but in fair condition with the remains of two patent sails hanging on.

Although these are now removed, by 1994 the fantail had been restored and the boat-shaped cap repaired and freshly painted white. Its internal gear connecting to an external scoop wheel survives.

Walpole's Mill, Halvergate
(TG 463070)   Y37
The Fleet, Halvergate Marshes

Almost opposite Howard's Mill on the south bank of The Fleet, Wailes (24) lists another mill draining Lord's Level on Halvergate Marshes, shown by Faden in 1797 as 'Drain W. Mill' and by the 1837 1" OS map as 'Mill' with the usual X symbol. The 1839 Halvergate Tithe Map shows a circular building here, apportioned in 1842 as (385) 'Mill and Garden', where later the 1884 6" OS map shows 'Windmill (Pumping)'.

Lockgate Mill, Breydon Water
TG 480071   Y38
River Yare, Acle Marshes

1837 OS 1" sheet 47:
'Mill' X

1841 Freethorpe Tithe Map: X   1841 Apportionment: (130) 'Mill Yards and Drain'

1884 OS 6" sheet 78 NW: 'Lockgate Windmill (Pumping)'

Listed grade II 839-1/2/201

Wailes: 6
Smith: 10

photo: evelyn simak 2011

Draining the southern edge of Freethorpe parish's detached portion, this mill pumped its water directly over the sea wall into Breydon Water, just to the east of where The Fleet joins this stretch of estuary.

Wailes attributes the mill to millwright Thomas Smithdale of Norwich and later Acle. Smith has it without cap or sails, but retaining most of the gear inside and a large scoop wheel outside.

Now with a temporary aluminium protective cap the gear, including windshaft, brakewheel, upright drive shaft, crown wheel, pit wheel and horizontal drive shaft, all appears to have survived by virtue of its being cast iron.

River Waveney Drainage Mills

We have now reached the end of the River Yare and will now finally turn our attention to the River Waveney. Starting at Beccles and taking just two stages, we again terminate here on Breydon Water, after the river forms the county boundary between Suffolk and Norfolk for many miles of its course.

Again our first site at Barsham (W1) lies too far west of Beccles to show here.

2.5 Upper River Waveney Drainage Windmills

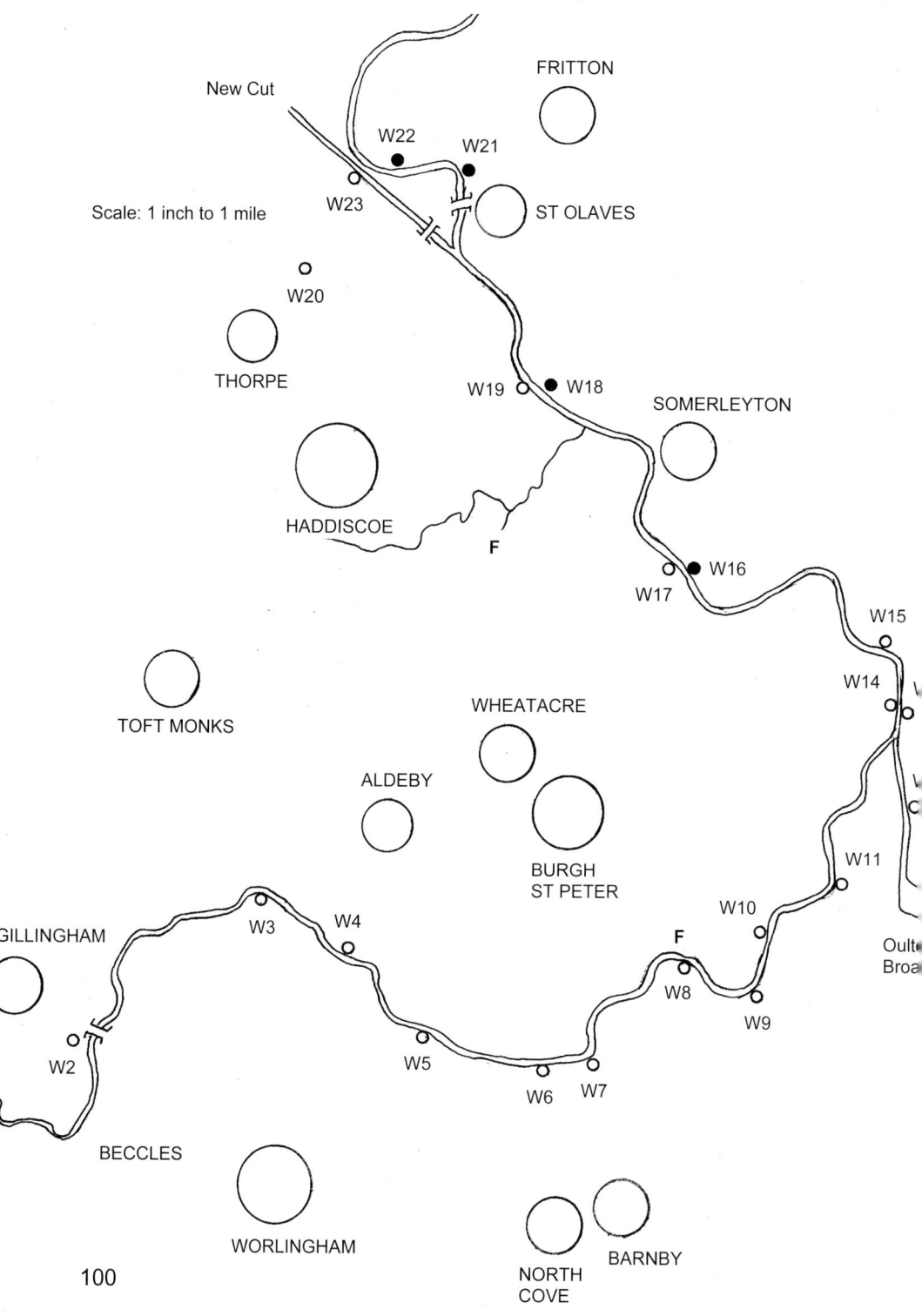

Railway Mill, Beccles
(TM 437927)   W3
River Waveney, Beccles Marshes

1838 OS 1" sheet 46: X

1840 Beccles Tithe Map: marshes area not shown

1884 OS 6" sheet 99 NE: 'Engine House'

Wailes —
Smith: —

Near where a now dismantled railway crossed the Waveney and shown as 'Engine House' on the 1884 6" OS map, the early 19th Century wind-powered mill at this site had been replaced by a steam engine in 1867, as the plaque attached to this brick building attests.

photo: polystar 2016

Barsham Hall Mill, Barsham
(TM 399911)   W1
River Waveney, Barsham Marshes

Adjoining a dyke off the River Waveney below the remains of the moated site of Barsham Hall, this drainage mill is the furthest west to be found up the Waveney valley (not counting the water supply pumping mill at Starston, TM 232843).

'Mill Marsh' is listed in the 1839 Barsham Tithe apportionment and the mill was shown as X on the 1838 1" OS map and as 'Windmill (Pumping)' on the 1884 6" OS map sheet 99 SW.

Bridge Mill, Gillingham
(TM 419912)   W2
River Waveney, Gillingham Marsh

Up a short dyke on the Norfolk side of the Waveney, between the old and new Beccles Bridges Faden shows in 1797 a 'Drain Mill' serving Gillingham Marsh, also shown on Hodskinson's 1783 map of Suffolk. This too appeared as X on the 1838 1" OS sheet 46, but not on the 1884 6" OS map sheet 99 SE.

The 1839 Gillingham Tithe Map also has X here, apportioned in 1845 (19) as 'Mill & Pightle'.

Four Upper Waveney Mills

River Waveney, Various Marshes

Aldeby Hall Mill, Aldeby
(TM 444922) W4
River Waveney, Priory Marsh

Moving downstream from Beccles the next drainage mill was on the northern bank of the River Waveney below Aldeby Hall, the site of a Benedictine Priory.

The mill appeared with the symbol X on the 1837 1" OS map, sheet 47, but had been replaced by 'Engine House' by the time the 6" map was published in 1884.

A similar fate came to many of the mills in this area, presumably because of the easy access to coal from the north via Lowestoft.

North Cove Mill, North Cove
(TM 465911) W6
River Waveney, North Cove Marsh

About a mile east of Worlingham Mill, yet another was shown as X on the 1837 1" OS sheet 47, but on the north bank of the River Waveney, presumably a cartographic error. It also appears as 'Marsh Mill' X on the 1848 North Cove Tithe map and later on the 1884 6" map sheet 99 SE on the correct side, annotated appropriately 'North Cove Mill (Pumping)' with a circular ground plan adjoining a rectangular one.

Wailes (105) describes this one under his Broads Area Suffolk heading as a tower mill that drove a turbine pump.

Worlingham Mill, Worlingham
(TM 453914) W5
River Waveney, Worlingham Marsh

Worlingham village, now almost absorbed by the growth of Beccles, had its own area of grazing marshes on the south bank of the River Waveney along with its own mill to drain it shown as 'Marsh Mill' X on the 1840 Worlingham Tithe map.
Appearing as another symbol X on the 1837 1" OS map, it too was replaced by a steam engine that was shown as 'Worlingham Mill (Pumping)' on the 1884 6" OS map sheet 99 SE.

Black Mill, North Cove
(TM 470912) W7
River Waveney, North Cove Marsh

About half a mile beyond North Cove Mill, there was another at Six Mile Corner, not shown on the 1837 1" OS map.
This appears in 1884 on the 6" map as 'Black Mill (Pumping)' again with a circular building plan next to a rectangular one, so it was not always an Engine Shed.

Castle Mill, North Cove
(TM 479921) W8
River Waveney, Castle Marsh

1837 OS 1" sheet 47:
'Castle Mill' X

1848 North Cove Tithe Map:
— (damaged area)
1848 Apportionment: (288) 'Mill Marsh'

1884 OS 6" sheet 100 NW:
'Castle Mill (Pumping)'

Wailes: 100
Smith: —

photo: polystar 2016

Another early wind-powered mill, Castle Mill drained Castle Marsh in the east of North Cove parish. Wailes has it as driving a turbine pump, and by 1884 it too had been replaced by an engine, presumably inside this brick shed.

Hober Mill, Barnby
(TM 487918) W9
River Waveney, Holbon Marsh

Around the bend in the Waveney east of Castle Mill, was another shown as 'Hober Mill' X on the 1837 1" OS sheet 47. The 1847 Barnby Tithe map shows a circular building apportioned as (281) 'Mill and Yard'.
This drained the marshes north of Barnby village, but not for very long as it was not shown on the 1884 6" map, although a surviving semi-circular ditch does appear there.

Chettleburgh's Mill, Burgh St Peter
(TM 488924) W10
River Waveney, Short Dam Level

Not on the 1837 1" OS map, Wailes (103) records Chettleburgh's Mill with a tower and scoop wheel draining Short Dam Level north of the Waveney, attributing it to millwright Martin. This site is confirmed by the 1842 Burgh St Peter Tithe Map with a circular building apportioned in 1843 as (413) 'Mill & Yard'. The 1884 6" OS map sheet 100 NW also has a circular building plan shown as 'Windmill (Pumping)'. In 1797 Faden had shown a 'Drain Mill' further west up a short dyke at TG 479924.

Share Mill, Carlton Colville
(TM 494927)   W11
River Waveney, Share Marsh

1837 OS 1" sheet 47:
'Share Mill' X

1842 Carlton Colville Tithe Map: —

1884 OS 6" sheet 100 NW:
'Share Mill (Pumping)'

Wailes: —
Smith: —

photo: polystar 2016

There was definitely a wind-powered drainage mill at this site in 1837, but Flint notes it had been replaced with an engine by 1883. Obviously it has been replaced again more recently with this rather unsightly green metal box containing a pump.

Oulton Broad Mill, Oulton Broad
(TM 517924 or 516928)
River Waveney, Oulton Marshes

The southernmost of these two potential sites, now a boatyard, is given by Wailes (106), where he places a tower mill with a scoop wheel, however his map references are not always reliable.

The northern alternative on the northern bank of Oulton Broad is given by Flint, also as a tower mill, but there appears to be no supporting map evidence for either?

Skepper's Mill, Oulton
(TM 502937)   W12
River Waveney, Oulton Marshes

A drainage mill served the southern part of Oulton Marshes on the eastern bank of Oulton Dyke, that connects Oulton Broad to the Waveney.
It was shown as 'Skepper's Mill' on the 1837 1" OS map and as (42) 'Marsh Mill and Yard' on the 1843 Oulton Tithe Apportionment. Flint has it as a trestle type mill, noting that there were five brick piers left standing, not found on a recent visit.

Somerleyton Mill, Somerleyton
TM 480959 W16
River Waveney, Somerleyton Marshes

1837 OS 1" sheet 47: X

1843 Somerleyton Tithe Map: circular building plan
1844 Apportionment: (297) 'Drainage Mill and Yard'

1884 OS 6" sheet 90 SW: 'Windmill (Pumping)'

Wailes: —
Smith: 75

photo: polystar 2016

South of Somerleyton village, the marshes lie on the north-east side of the River Waveney, which here flows roughly north-westwards towards the New Cut. Here can be found the first mill remains of any consequence as one passes downstream from Beccles.

It was a tower mill of some age, but is now reduced to a mere stump, little more than a circular brick hut with a conical corrugated iron roof. Flint notes that it drove a turbine pump. A prominent local landmark can be seen in the background: the chimney of the engine house that replaced Black Mill, Wheatacre.

> **Arnold's Mill, Oulton**
> (TM 501946) W13
> *River Waveney, Oulton Marshes*
>
> The northern end of Oulton Marshes was drained by Arnold's Mill on the east bank of the Waveney just past the junction with Oulton Dyke. It appeared on Hodskinson's 1783 map and early OS maps as X in 1837 and 'Windmill (Pumping)' in 1884. The Oulton Tithe Apportionment has it as (107) 'Marsh Mill and Yard'.
> Flint says it ceased working in 1903, but that a fifteen foot high stump remained, which was demolished in 2000.

Black Mill, Wheatacre
(TM 478959)   W17
River Waveney, Wheatacre Marshes

1837 OS 1" sheet 47: 'Black Mill' X

1840 Wheatacre Tithe Map: X   1841 Apportionment: (19a) 'Drainage Mill Yard'

1884 OS 6" sheet 90 SW: 'Black Mill (Pumping)'

Wailes: —
Smith: Some Old Sites

Almost opposite Somerleyton Mill on the south-west bank of the River Waveney, there was another tower mill that drained Wheatacre Marshes on the Norfolk side as shown by the early maps. Replaced by a steam engine with a tall octagonal chimney that remains to this day, it became known by passing wherrymen as the 'Barber's Pole'.

photo: robert newell 2011

Smith includes it in his 'Some Old Sites' list, but Wailes seems to have missed this one, although he does include (18) Powell's Mill (W23) at an incorrect map reference TM 447961, further west near the village of Haddiscoe.

Interestingly in 1797 Faden showed a 'Drain W. Mill' not so far west in Aldeby at TM 460961, now long gone.

Burgh Mill, Burgh St Peter
(TM 500947)   W14
River Waveney, Burgh Marshes

Almost opposite Arnold's Mill on the west bank of the River Waveney, there was a mill that drained Burgh Marshes to the west.

It appeared on the 1837 1" OS map as 'Burgh Mill' with the usual X symbol. The later 1884 6" OS map shows 'Burgh Windmill (Pumping)' here and Smith includes it in his 'Some Old Sites' list, noting 'ruined portions of lower walls & rubble pile'.

Herringfleet Mill, Herringfleet
TM 465976　　　W18
River Waveney, Herringfleet Marshes

1837 OS 1" sheet 47: —

1849 Herringfleet Tithe Map: circular building plan
1849 Apportionment: —

1884 OS 6" sheet 90 SW: 'Windmill (Disused)'

Listed grade II* 1/14

Wailes: 108
Smith: 74

Given a higher grade II* listing as the last full size smock mill in the Broads area, this one is octagonal in plan and clad in black weatherboarding. It drained Herringfleet Marshes on the Suffolk side of the River Waveney. Of early 19th Century date, it has been restored to working order by Suffolk County Council.

photo: ashley dace 2010

The mill has common sails and an external scoop wheel, with the driving shaft inscribed 'W.T England. Millwright. Yarmouth 1910', presumably a rebuild date.
It seems likely that the X symbol shown on the 1837 1" OS map on the opposite bank here (TM 463975) is an earlier mill (W19) since Faden in 1797 shows a 'Hadiscoe Drain Mill' there and the 1841 Haddiscoe Tithe Map shows a semi-circular enclosure apportioned (111) as 'Mill Yard'.

Blundeston Mill, Blundeston
(TM 499952)　　　W15
River Waveney, Blundeston Marshes

About half a mile due north of Burgh Mill, but on the opposite Suffolk bank of the Waveney, which has by now turned north-westwards, there was a mill draining Blundeston Marshes, shown as (29) 'Mill and Yard' on the 1842 Tithe Apportionment.
It appeared with the usual X symbol on the 1837 1" OS map and later in 1884 on the 6" as 'Windmill (Pumping)'.

St. Olave's Mill, Fritton
TM 457997　　　W21
River Waveney, Fritton Marshes

1837 OS 1" sheet 47: —

1849 Herringfleet Tithe Map: circular building plan
1849 Apportionment: —

1884 OS 6" sheet 89 NE: 'Windmill (Pumping)'

Listed grade II* 10/23

Wailes: 110
Smith: 21

photo: polystar 1994 S4

In 1910 this small square section smock type mill replaced an earlier octagonal smock mill at this site, which had been demolished in 1898 according to Flint. Wailes (109) suggests a further mill nearby over St Olave's Bridge at TG 455995, for which there is no map evidence.

Sometimes known as Priory Mill, Wailes also calls it Pony Mill, which might be the result of poor transcription. He attributes the current mill driving an internal scoop wheel to millwright W.T. England. It was restored to its current working condition in 1980. The stocks, sails and fantail date from the restoration as Smith's photograph shows it without these, but all the internal machinery is intact and cast iron except for the timber upright shaft.

Thorpe Mill, Thorpe by Haddiscoe
(TG 441987)　　　W20
River Waveney, Thorpe Marshes

Thorpe Marshes to the west of the Waveney end of New Cut were drained by 'Thorpe Drain Mill' according to Faden in 1797. Also shown as 'Old Mill' on the 1837 1" OS map, it appears as a symbol X on the 1839 Thorpe Tithe Map, apportioned in 1840 as (114) 'Mill and Yard'. It was however no longer there by the time the 6" OS map sheet 89 NE was published in 1884.

Fritton Marsh Mill, Fritton
TM 450998   W22
River Waveney, Fritton Marshes

1837 OS 1" sheet 47: X

1838 Fritton Tithe Map: X
1840 Apportionment: (65) 'Great Mill Marsh' adjoining

1884 OS 6" sheet 89 NE: 'Windmill (Pumping)'

Listed grade II 10/22

Wailes: 104
Smith: 20

About half a mile west of St. Olave's Mill on the north bank of the here westwards flowing River Waveney, this mill drained the southern part of Fritton Marshes, formerly part of Suffolk. Wailes attributes it to millwright Martin. In Smith's eyes it was another derelict tarred red brick tower without cap or sails.

photo: pierre terre 2008

It does retain its windshaft, brake-wheel and the timber upright shaft and a 7'2" diameter pit wheel that drove an external scoop wheel, albeit all now under a novel corrugated iron cap. The listing takes account of the survival of the gear and it also has an external engine shed.

Powell's Mill, Raveningham (det)
(TM 447997)   W23
River Waveney, Thurlton Marshes

When New Cut arrived in 1833 'Raveningham Drain Mill' shown by Faden in 1797 on the south bank of the Waveney (TG 447998), was severed from the land it drained. A replacement was built adjoining New Cut, listed in the 1841 Raveningham Tithe Apportionment as (56) 'Cot Mill & Garden', and shown as 'Windmill (Pumping)' on the 1884 6" OS. Wailes picks up on the dual nature of this site with Tom Cook's Mill (38) at Loddon (sic) (Detached), converted by W.T. England and Powell's Mill (18) at Haddiscoe Marshes rebuilt by Barnes.

Lower River Waveney Drainage Windmills

Scale: 1 inch to 1 mile

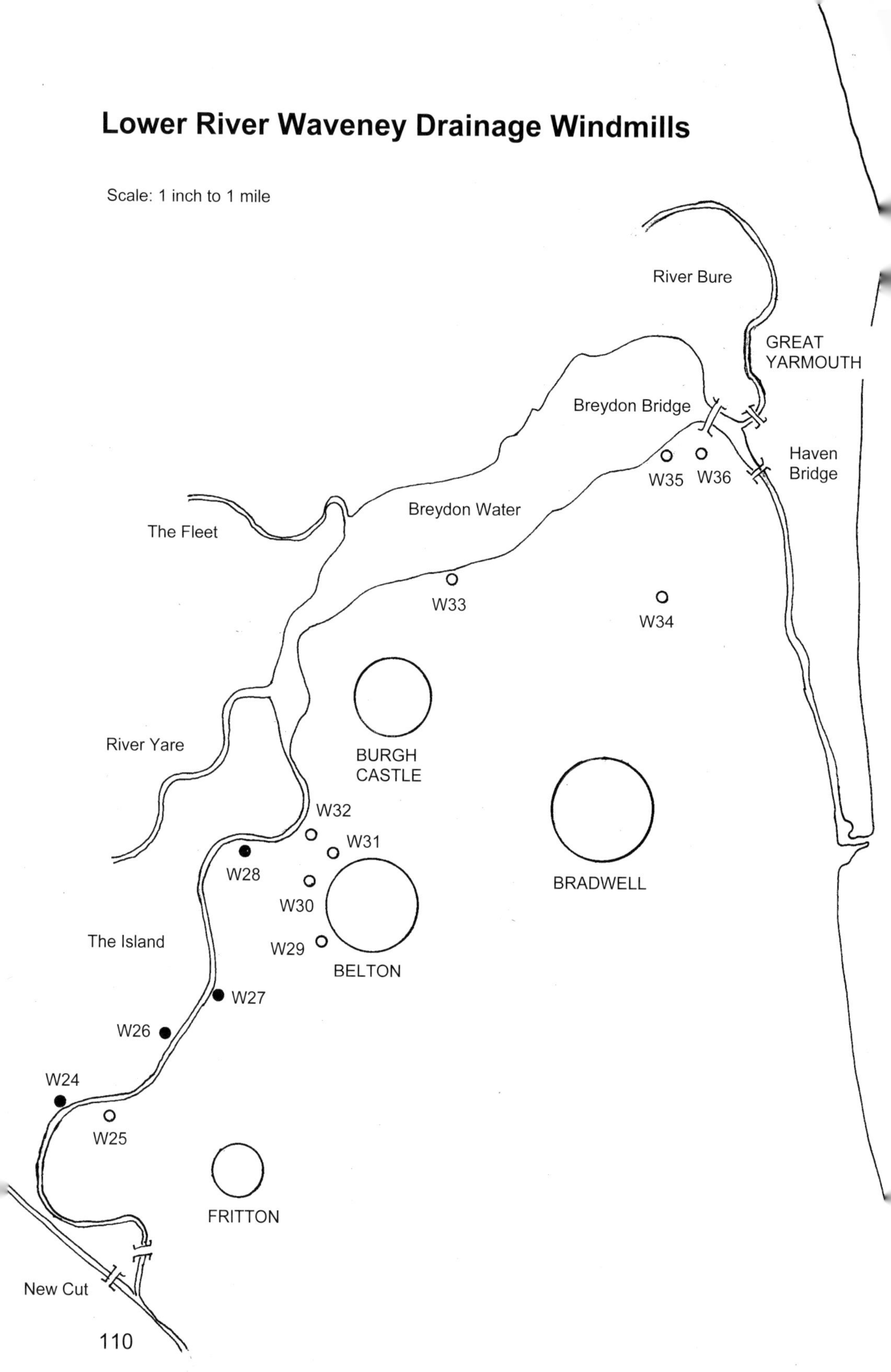

- River Bure
- GREAT YARMOUTH
- Breydon Bridge
- W35 W36
- Haven Bridge
- Breydon Water
- The Fleet
- W33
- W34
- River Yare
- BURGH CASTLE
- W32
- W31
- W28
- W30
- BRADWELL
- The Island
- W29
- BELTON
- W27
- W26
- W24
- W25
- FRITTON
- New Cut

110

Toft Monks Mill, Haddiscoe Island
TG 448009   W24
River Waveney, Toft Monks Marsh

1797 Faden: 'Drain W. Mill'

1837 OS 1" sheet 47: X

1840 Toft Monks Tithe Map: X   1842 Apportionment: (299) 'Mill & Yard'

1884 OS 6" sheet 89 NE: 'Windmill (Pumping)'

Listed grade II 3/10

Wailes: 12
Smith: 17

photo: evelyn simak 2010

Although in a detached portion of Toft Monks parish, Wailes describes this drainage mill on the north-west bank of the River Waveney alongside Haddiscoe Island, as Rose's Mill on Chedgrave Marsh. He states 'No notes' and gives no map reference.

Smith certainly includes it describing a derelict red brick tower in poor condition without a cap, but still retaining three ruined patent sails, its windshaft and internal gear with upright shaft that drove an external scoop wheel, reduced to a 'skeleton'.

The gear is now protected by a replica weather-boarded cap, originally white but now black, and there is a modern but derelict pump-house adjacent.

Fritton Warren Mill, Fritton
(TG 453008)   W25
River Waveney, Fritton Marshes

The northern part of Fritton Marshes was drained in latter days by a mill erected in 1910 and demolished around 1948 according to Flint, so there is no reference on any old maps.

Its nearness to 'Skeleton Wood' suggests an open frame, but Wailes (107) describes it as a smock mill by W.T. England, driving a turbine pump.

Pettingell's Mill, Haddiscoe Island
TG 459016   W26
River Waveney, Chedgrave Marshes

1783 Hodskinson's: X

1837 OS 1" sheet 47: X

1840 Toft Monks Tithe Map: X   1842 Apportionment: (315) 'Marsh House, Mill &c'

1885 OS 6" sheet 77 SE: 'Draining Pump'

Listed grade II 4/12

Wailes: 74
Smith: 16

A mile downstream from Toft Monks Mill, the next one on The Island, draining Chedgrave Marshes into the River Waveney is near Seven Mile House. Many farmhouses along the Rivers Bure, Yare and Waveney are so named for their distances upstream from Great Yarmouth.

photo: polystar 1994 S7

Wailes names this one as Pettingell's Mill at Toft Monks, the parish this detached portion of The Island belongs to. Smith lists it as a preserved (rather than derelict) tarred red brick tower only. It has a corrugated iron roof covering and scoop wheel housing.

A little overgrown in 1994, the mill sits back a bit from the river, its scoop wheel delivering the water into a short connecting dyke.

Three Mills, Belton   W29, 30 & 31
(TG 477026, 474032 & 477035)
River Waveney, Belton Marshes

Well away from the River Waveney at the back edge of Belton Marshes below the higher ground of Belton village, there were three drainage mills all shown on the 1884 6" OS map sheet 78 SW as 'Windmill (Pumping)'.

None of these had appeared on the earlier maps, and not on more recent maps so they were fairly short-lived.

Caldecott Mill, Fritton
TG 464021   W27
River Waveney, Belton Marshes

1783 Hodskinson's: X

1837 OS 1" sheet 47: X

1838 Fritton Tithe Map: X
1840 Apportionment: (3)/(4) 'Great / Little Mill Rand' adjoining

1884 OS 6" sheet 78 SW: 'Windmill (Pumping)'

Listed grade II 7/21

Wailes: 102
Smith: 19

photo: evelyn simak 2015

Still just in the parish of Fritton, but draining the southern part of Belton Marshes to its north, this mill was listed by Wailes as in Belton. Smith meanwhile lists it as at Bell Hill, a promontory to the south.

Like Pettingell's Mill half a mile to the south-west, it is set back from the main river frontage a little, draining into a short dyke. The actual river here is effectively set within a secondary flood plain between the sea walls either side.

Another derelict tarred red brick tower, without cap or sails, the internal gear is protected by a temporary roof since a fire in 1991. The pit wheel drove a shaft to an external pump in its own housing.

Mill, Burgh Castle
(TG 474037)   W32
River Waveney, Belton Marshes

Just north of the three Belton Mills, but just over the parish boundary into Burgh Castle and a little nearer the River Waveney, there was another mill.

This similarly did not appear on the early maps, but is shown just on the 1884 6" OS map sheet 78 SW as 'Windmill (Pumping)'.

Black Mill, Belton
TG 468034   W28
River Waveney, Belton Marshes

1837 OS 1" sheet 47: X

1838 Belton Tithe Map: —
1840 Apportionment: —

1884 OS 6" sheet 78 SW:
'Windmill (Pumping)'

Listed grade II 7/1

Wailes: 101
Smith: 8

photo: polystar 2016

Draining the northern part of Belton Marshes, this mill is the last with any appreciable remains that can be seen travelling down the River Waveney. It is relatively remote and about a quarter mile away from the course of the river itself.

Wailes has it as rebuilt in 1907 by millwright Hewitt of Berney Arms and it was further restored by Smithdale of Acle in 1973. Much of the internal gear has been re-used from previous builds and Smith describes a derelict red brick tower, but with an unusual faded green boat-shaped cap.

The convex curved batter, more than would be needed for visual correction, is curiously described in the listing as 'entatic'.

Mill, Bradwell
(TG 512062)   W34
River Waveney, Gapton Marshes

Yet another mill that appeared just on the 1884 6" OS map as 'Windmill (Pumping)' and nowhere else, was to be found on Gapton Marshes, Bradwell.

In a similar situation to the three Belton Mills, the site immediately adjoins modern development on the marsh edge about a mile south of the bank of Breydon Water.

Burgh Mill, Burgh Castle
(TG 489064)   W33
River Waveney, Burgh Castle Marshes

1837 OS 1" sheet 47: 'Burgh Mill' X

1843 Burgh Castle Tithe Map: X 1843 Apportionment (47): 'Mill Yard & Outfall'

1884 OS 6" sheet 78 NW: 'Windmill (Pumping)'

Wailes: —
Smith: —

photo: polystar 2016

Down stream from Black Mill, beyond the enormous Roman fort known as Burgh Castle, this small red brick pump-house is something of a return to what we saw in the upper reaches of the River Waveney.

The original wind-powered mill here appeared as such on the early maps and survived until at least 1884. It drained Burgh Castle Marshes on the southern shore of Breydon Water, below the confluence of the Rivers Yare and Waveney.

The pumps here are now electrically driven and it is still shown on modern OS maps as 'Pump House'.

Two Mills, Cobholm Island
(TG 511077 & 515077)   W35 & W36
River Waveney, Cobholm Marsh

Finally on the south bank near the mouth of Breydon Water, there were two drainage mills at Cobholm Island. Both were shown as X on Hodskinson's 1783 map of Suffolk along with another one opposite in Great Yarmouth. Faden also showed the pair in 1797.

The 1837 1" OS sheet shows 'Mills' here with two X symbols, but only the western one appears later on the 1886 6" map sheet 78 NE as 'Draining Pump'. Wailes (15) lists this one as 'Jiber's Mill', but with an incorrect map reference.

Four mills improved since 1994

Norton Marsh Mill Y18 (p.86)

Langley Detached Mill Y29 (p.93)

Fritton Marsh Mill W22 (p.109)

Toft Monks Mill W24 (p.111)

3.1 Summary

So we come to the end of our gazetteer and can see that we have examined 155 sites where wind-powered mills pumped water off the marshes into the adjoining rivers. Of these 74 have left visible remains which means that 81 have been lost, so overall we are looking at a survival rate of about one half. More specifically about two thirds of the mills have survived on the Rivers Thurne and Bure, just less than half on the Rivers Ant and Yare, but less than a quarter along the River Waveney, the least remote of the rivers, where early on they were replaced with steam engines and eventually diesel or electricity took over.

Of the 74 survivors, one is a scheduled monument and some 50 are statutorily protected by listing. Of these 36 are grade II, 13 are given the higher grade II* and one (St Benet's Abbey) the highest grade I listing and scheduled monument status too. A number of the listed ones are in poor condition and consequently appear on the 'Buildings at Risk' register. Whilst the 23 unlisted ones are perhaps most at risk they are given some protection by being included in a 'Local List' produced by The Broads Authority, which also includes the five surviving pumping stations that are not listed.

Unlike other Buildings at Risk, the problem with mills is one of finding a suitable beneficial use which will not erase historic evidence as a residential conversion might. Whilst the removal of internal gear to facilitate conversion does save the tower as a landmark within the wider historic environment, it leaves little of actual interest for the building historian or mill enthusiast.

It is hoped that the documentation of these mills as presented here will be the foundation for further work in years to come, providing a definitive listing of what there was and what we still have removing the errors that crept into previous accounts. Those accounts have certainly been useful starting points, but Wailes seems to have missed some six sites (Hickling Broad, Catfield Middle Marsh, Clayrack, Belaugh Old Hall, Boyce's Dyke and Somerleyton), some more surprisingly than others. Smith did include most of these, but seems to have left out three sites altogether (Moy's, Catfield Middle Marsh and understandably Clayrack whilst being moved to its new location).

The present work has been written to be as comprehensive as possible, re-examining the early map evidence and taking on board the valiant efforts of the earlier accounts. The new river by river numbering system should help here and hopefully be flexible enough to allow the future inclusion of anything unwittingly missed at this point in time. Hopefully the listing along each river in turn will make the information presented here more useful and appealing to those on boating holidays and thus promote a wider understanding of this subject.

3.2 References and Bibliography

Alderton, D. & Booker, J. 1980 *The Industrial Archaeology of East Anglia* Batsford

Apling, H. 1984 *Norfolk Corn Mills* Norfolk Windmills Trust

Armstrong, P. 1975 *The Changing Landscape* Terence Dalton

Chatwin, C.P. 1961 *East Anglia and Adjoining Areas* British Regional Geology HMSO

Clifford, S. & King, A. (eds) 1993 *Local Distinctiveness* Common Ground

Dolman, P.C.J. 1978 *Windmills in Suffolk* Suffolk Mills Group

Flint, B. 1979 *Suffolk Windmills* Boydell

Malster, R. 2003 *The Norfolk & Suffolk Broads* Phillimore

Norfolk Windmills Trust 1982 *Windmills to Visit*

Ordnance Survey 1838 *Cromer* 1" Reprint 38 David & Charles

Ordnance Survey 1838 *Norwich* 1" Reprint 46 David & Charles

Ordnance Survey 1837 *Great Yarmouth* 1" Reprint 47 David & Charles

Ordnance Survey 1984 *Norwich and The Broads* Landranger 134 HMSO

Ordnance Survey 2015 *The Broads* Explorer 40 HMSO

Smith, A.C. 1978 *Drainage Windmills of the Norfolk Marshes* Stevenage Museum Publications

Smith, A.C. 1982 *Corn Windmills in Norfolk* Stevenage Museum Publications

Wade-Martins, P. (ed) 1994 *An Historical Atlas of Norfolk* Norfolk Museums

Wailes, R. 1957 *Norfolk Windmills: Part II, Drainage and Pumping Mills Including Those of Suffolk* Transactions of the Newcomen Society Vol. XXX

3.3 Acknowledgements

My thanks must go to Robert Malster for some early input at a Suffolk Industrial Archaeology Society meeting at which I first aired my newly transferred digital slides and outlined this project, and also to Mark Barnard of Suffolk Mills Group for some early discussions on the subject.

The book may not be as 'in depth' as they would have liked since my intention has simply been to provide a comprehensive survey of what was and what remains. This will hopefully now be a starting point for further research by those who do know the subject in greater detail than myself and also an encouragement for the layman, who as yet knows relatively little, to get involved.

I would like to thank the helpful staff members at Suffolk Record Office, Norfolk Record Office and the Norwich Heritage Centre for their production of so many maps and I am indebted to the several 'Creative Commons' photographers whose recent efforts to document these buildings photographically have saved me many a mile of walking.

I must also thank Will Burchnall and Ben Hogg of the Broads Authority for their helpful advice, encouragement and an overview of the proofs. Finally and most importantly, my dear departed dedicatee Dil Andrews, whose message follows:

Enjoy your boating!

Other Books of Local Interest

The Toll-houses of Norfolk
Patrick Taylor   2009   £7.95
ISBN 978 1 907154 02 7   iv+76pp
Polystar Press

Research interrupted by the Norwich Library fire finally resumed and brought to publication.

Essentially the same format as the Suffolk volume: history of the turnpike roads, detailed gazetteer of the county plus an appendix on the impostors.

"timely and important records of those that survive and also those - sadly the majority - that have been lost"
Norfolk Industrial Archaeology Soc.

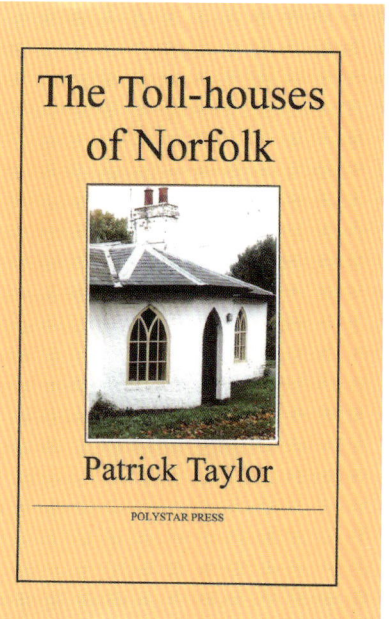

Timber Circles in the East
Patrick Taylor   2015   £8.95
ISBN 978 1 907154 60 7   iv+116pp
Polystar Press

The first published study of the east of England's timber circles as discovered thus far, it takes them as the eastern counterpart of the west's stone circles and looks at them in the light of Alexander Thom's work on stone circles, discovering similar geometries in their layouts and similar alignments to the extreme rising and setting points of the sun and moon.

"nicely produced, literate and informative...excellent plans and photographs...small but well-written book packs a lot into its 116 pages..."
R.I.L.K.O. Journal